# EACHER GUIDE

Includes St...
Worksh...

## 7th – 12th Grade

Science

Quizzes

MW00335078

# General Science 2: Survey of Geology & Archaeology

**Master Books® Curriculum**

**Master Books Creative Team:**

**Author:** Emil Silvestru,
Dr. Gary Parker, David Down,
Dr. John D. Morris

**Editor:** Craig Froman

**Design:** Terry White

**Cover Design:**
Diana Bogardus

**Copy Editors:**
Judy Lewis, Willow Meek

**Curriculum Review:**
Kristen Pratt, Laura Welch
Diana Bogardus

First printing: March 2016
Sixth printing: February 2021

Master Books®, P.O. Box 726, Green Forest, AR 72638
Master Books® is a division of the New Leaf Publishing Group, Inc.

ISBN: 978-0-89051-966-0
ISBN: 978-1-61458-558-9 (digital)

Unless otherwise noted, Scripture quotations are from the New King James Version.
Copyright © 1982 by Thomas Nelson, Inc. Used by permission. All rights reserved.

**Printed in the United States of America**

Please visit our website for other great titles:
www.masterbooks.com

**Author Bios: Emil Silvestru** is a prospecting and exploration geologist. He is a cave expert born in Transylvania, Romania, and is currently a writer, researcher, and speaker. **Dr. Gary Parker** lectures worldwide for both ICR and Answers in Genesis, and has written five science textbooks and numerous creation books. **David Down** was a field archaeologist for over four decades, excavating regularly in Israel, and offers special insight into the connections of archaeology and biblical history. **Dr. John D. Morris** is president emeritus of the Institute for Creation Research and currently travels and speaks on the topic of creation science.

Your reputation as a publisher is stellar. It is a blessing knowing anything I purchase from you is going to be worth every penny!

—Cheri ★★★★★

Last year we found Master Books and it has made a HUGE difference.

—Melanie ★★★★★

We love Master Books and the way it's set up for easy planning!

—Melissa ★★★★★

You have done a great job. MASTER BOOKS ROCKS!

—Stephanie ★★★★★

Physically high-quality, Biblically faithful, and well-written.

—Danika ★★★★★

Best books ever. Their illustrations are captivating and content amazing!

—Kathy ★★★★★

# Affordable
# Flexible
# Faith Building

# Table of Contents

# Using This Teacher Guide

**Features:** The suggested weekly schedule enclosed has easy-to-manage lessons that guide the reading, worksheets, and all assessments. The pages of this guide are perforated and three-hole punched so materials are easy to tear out, hand out, grade, and store. Teachers are encouraged to adjust the schedule and materials needed in order to best work within their unique educational program.

**Dig into Science!** And not just figuratively! Enjoy exploring ancient cities and cultures, types of rocks and minerals, the formation of caves, some of the best places to find fossils, and much more. Through this course, students will learn how the earth's features support the biblical account of creation, corruption, catastrophe, and Christ — and how evolution cannot possibly be true. The course includes worksheets, quizzes, and tests, as well as activities such as excavation exercises, research, collecting, and field trips.

| | |
|---|---|
| Approximately 30 to 45 minutes per lesson, five days a week | |
| Includes answer keys for worksheets, quizzes, and tests | |
| Worksheets for each section | |
| Quizzes and tests are included to help reinforce learning and provide assessment opportunities | |
| Designed for grades 7 to 12 in a one-year course to earn 1 science credit | |

# Course Description

This is the suggested course sequence that allows two core areas of science to be studied per semester. You can change the sequence of the semesters per the needs or interests of your students; materials within each semester are independent of one another to allow flexibility. The following is how this teacher guide has been structured for the year of study. The first quarter covers archaeology and takes students on an exciting exploration of history and ancient cultures. The second quarter covers geology, guiding students to see what really carved the Grand Canyon, how thick the earth's crust is, why the earth is unique for life, and the varied features of the earth's surface. The third quarter focuses on caves, exploring deep into the hidden wonders beneath the surface as cave expert Dr. Emil Silvestru takes students on this illuminating and educational journey. The fourth quarter concludes the year with a study of fossils and their origins, as well as how to collect them.

# Course Objectives

Students completing this course will:

- Evaluate how archaeologists know what life was like in the past, some of the difficulties in dating ancient artifacts, and how the brilliance of ancient cultures demonstrates God's creation

- Discover the beautiful, thriving ecology, unique animals, and fragile balance of this little-seen ecosystem in caves from around the globe

- Review how the Flood affected fossil formation, the difference between evolutionists' and creationists' views on fossils, and the "four C's" of biblical creation

- Investigate what really carved the Grand Canyon, how thick the earth's crust is, why the earth is unique for life, and the varied features of the earth's surface

- Explore the origin of fossils, how to start your own fossil collection, what kinds of fossils can be commonly found, the age of fossils, and how scientists find and preserve fossils

- Learn both the techniques of the archaeologist and the accounts of some of the richest discoveries of the Middle East that demonstrate the accuracy and historicity of the Bible

- Identify a creationary model for how caves form, a history of how caves have been used by humans for shelter and worship, and how old caves really are

**Special Note:** High school students who take the course are expected to do a majority of the activities. The activities can be modified based on student interests and creativity, but should reflect an understanding of the core concepts being learned.

# Introduction

## How worksheets are designed:

**Words to Know/Define:** As you read through the assigned readings, write down the definitions and meanings of these significant words. Most can be found in the glossary in the back of this teacher guide.

**Questions/Short Answer:** Seek out and note the answers to these questions, knowing you can review these prior to taking the unit quizzes and tests.

**Activities:** These are optional activities that help reinforce ideas taught in the books.

## Grading process:

A teacher can grade assignments daily or weekly and keep track of this in their files. Worksheet answers are available in the back of the book. You many use the following standard system for grading (90-100 = A, 80-89 = B, 70-79 = C, 60-69 = D, below 60 = F), or you may choose to create your own grading system.

# First Semester Suggested Daily Schedule

| Date | Day | Assignment | Due Date | ✓ | Grade |
|------|-----|------------|----------|---|-------|
| | | First Semester-First Quarter — *The Archaeology Book* | | | |
| **Week 1** | Day 1 | Read Pages 6–19 • *The Archaeology Book* • (AB) | | | |
| | Day 2 | What Archaeology Is All About - Words to Know<br>Ch 1: Worksheet 1<br>Pages 17–18 • *Teacher Guide* (TG) | | | |
| | Day 3 | What Archaeology Is All About - Questions<br>Ch 1: Worksheet 1 • Page 18 • (TG) | | | |
| | Day 4 | What Archaeology Is All About - Activities<br>Ch 1: Worksheet 1 • Page 18 • (TG) | | | |
| | Day 5 | Read Pages 20–29 • (AB) | | | |
| **Week 2** | Day 6 | Land of Egypt - Words to Know<br>Ch 2: Worksheet 1 • Page 19 • (TG) | | | |
| | Day 7 | Land of Egypt - Questions<br>Ch 2: Worksheet 1 • Pages 19–20 • (TG) | | | |
| | Day 8 | Land of Egypt - Activities<br>Ch 2: Worksheet 1 • Page 20 • (TG) | | | |
| | Day 9 | *Archaeology Book* Chapters 1–2 Study Day | | | |
| | Day 10 | ***Archaeology Book* Ch 1–2 Quiz 1** • Pages 99–100 • (TG) | | | |
| **Week 3** | Day 11 | Read Pages 30–35 • (AB) | | | |
| | Day 12 | The Hittites - Words to Know, Questions<br>Ch 3: Worksheet 1 • Pages 21–22 • (TG) | | | |
| | Day 13 | The Hittites - Activities<br>Ch 3: Worksheet 1 • Page 22 • (TG) | | | |
| | Day 14 | Read Pages 36–41 • (AB) | | | |
| | Day 15 | Ur of the Chaldees - Words to Know, Questions<br>Ch 4: Worksheet 1 • Pages 23–24 • (TG) | | | |
| **Week 4** | Day 16 | Ur of the Chaldees - Activities<br>Ch 4: Worksheet 1 • Page 24 • (TG) | | | |
| | Day 17 | Read Pages 42–45 • (AB) | | | |
| | Day 18 | Assyria - Words to Know, Questions<br>Ch 5: Worksheet 1 • Pages 25–26 • (TG) | | | |
| | Day 19 | Assyria - Activities<br>Ch 5: Worksheet 1 • Page 26 • (TG) | | | |
| | Day 20 | *Archaeology Book* Chapters 3–5 Study Day | | | |
| **Week 5** | Day 21 | ***Archaeology Book* Ch 3–5 Quiz 2** • Pages 101–102 • (TG) | | | |
| | Day 22 | Read Pages 46–51 • (AB) | | | |
| | Day 23 | Babylon: City of Gold - Words to Know, Questions<br>Ch 6: Worksheet 1 • Page 27 • (TG) | | | |
| | Day 24 | Babylon: City of Gold - Activities<br>Ch 6: Worksheet 1 • Page 28 • (TG) | | | |
| | Day 25 | Read Pages 52–59 • (AB) | | | |

| Date | Day | Assignment | Due Date | ✓ | Grade |
|------|-----|-----------|----------|---|-------|
| | Day 26 | Persia - Words to Know, Questions<br>Ch 7: Worksheet 1 • Pages 29–30 • (TG) | | | |
| | Day 27 | Persia - Activities<br>Ch 7: Worksheet 1 • Page 30 • (TG) | | | |
| Week 6 | Day 28 | Read Pages 60–69 • (AB) | | | |
| | Day 29 | Petra - Words to Know, Questions<br>Ch 8: Worksheet 1 • Pages 31–32 • (TG) | | | |
| | Day 30 | Petra - Activities<br>Ch 8: Worksheet 1 • Page 32 • (TG) | | | |
| | Day 31 | *Archaeology Book* Chapters 6–8 Study Day | | | |
| | Day 32 | **Archaeology Book Ch 6–8 Quiz 3** • Page 103 • (TG) | | | |
| Week 7 | Day 33 | Read Pages 70–77 • (AB) | | | |
| | Day 34 | The Phoenicians - Words to Know, Questions<br>Ch 9: Worksheet 1 • Pages 33–34 • (TG) | | | |
| | Day 35 | The Phoenicians - Activities<br>Ch 9: Worksheet 1 • Page 34 • (TG) | | | |
| | Day 36 | Read Pages 78–83 • (AB) | | | |
| | Day 37 | The Dead Sea Scrolls - Words to Know, Questions<br>Ch 10: Worksheet 1 • Pages 35–36 • (TG) | | | |
| Week 8 | Day 38 | The Dead Sea Scrolls - Activities<br>Ch 10: Worksheet 1 • Page 36 • (TG) | | | |
| | Day 39 | Read Pages 84–93 • (AB) | | | |
| | Day 40 | Israel - Words to Know, Questions<br>Ch 11: Worksheet 1 • Pages 37–38 • (TG) | | | |
| | Day 41 | Israel - Activities<br>Ch 11: Worksheet 1 • Page 38 • (TG) | | | |
| | Day 42 | *Archaeology Book* Chapters 9–11 Study Day | | | |
| Week 9 | Day 43 | **Archaeology Book Ch 9–11 Quiz 4** • Pages 105–106 • (TG) | | | |
| | Day 44 | *Archaeology Book* Chapters 1–11 Study Day | | | |
| | Day 45 | **Archaeology Book Test** • Pages 107–109 • (TG) | | | |
| First Semester-Second Quarter — **The Geology Book** | | | | | |
| | Day 46 | Read Pages 4–9 • *The Geology Book* • (GB)<br>Planet Earth - Words to Know, Questions<br>Intro & Ch 1: Worksheet 1 • Pages 41–42 • (TG) | | | |
| | Day 47 | Planet Earth - Activities<br>Intro & Ch 1: Worksheet 1 • Page 42 • (TG) | | | |
| Week 1 | Day 48 | Read Pages 10–19 • (GB) | | | |
| | Day 49 | The Ground We Stand Upon - Words to Know, Questions<br>Ch 2: Worksheet 1 • Pages 43–44 • (TG) | | | |
| | Day 50 | The Ground We Stand Upon - Activities<br>Ch 2: Worksheet 1 • Page 44 • (TG) | | | |

| Date | Day | Assignment | Due Date | ✓ | Grade |
|---|---|---|---|---|---|
| | Day 51 | Read Pages 20–27 • (GB) | | | |
| | Day 52 | The Earth's Surface - Words to Know<br>Ch 3: Worksheet 1 • Page 45 • (TG) | | | |
| Week 2 | Day 53 | The Earth's Surface - Questions<br>Ch 3: Worksheet 1 • Pages 45–46 • (TG) | | | |
| | Day 54 | The Earth's Surface - Activities<br>Ch 3: Worksheet 1 • Page 46 • (TG) | | | |
| | Day 55 | *Geology Book* Chapters 1–3 Study Day | | | |
| | Day 56 | ***Geology Book* Chapters 1–3 Quiz 1** • Pages 111–112 • (TG) | | | |
| | Day 57 | Read Pages 28–35 • (GB) | | | |
| Week 3 | Day 58 | Geological Processes and Rates - Words to Know<br>Ch 4: Worksheet 1 • Page 47 • (TG) | | | |
| | Day 59 | Geological Processes and Rates - Questions<br>Ch 4: Worksheet 1 • Pages 47–48 • (TG) | | | |
| | Day 60 | Geological Processes and Rates - Activities<br>Ch 4: Worksheet 1 • Page 48 • (TG) | | | |
| | Day 61 | Read Pages 36–41 • (GB) | | | |
| | Day 62 | Geological Processes and Rates - Words to Know<br>Ch 4: Worksheet 2 • Page 49 • (TG) | | | |
| Week 4 | Day 63 | Geological Processes and Rates - Questions<br>Ch 4: Worksheet 2 • Pages 49–50 • (TG) | | | |
| | Day 64 | Geological Processes and Rates - Activities<br>Ch 4: Worksheet 2 • Page 50 • (TG) | | | |
| | Day 65 | Read Pages 42–48 • (GB) | | | |
| | Day 66 | Geological Processes and Rates - Words to Know<br>Ch 4: Worksheet 3 • Page 51 • (TG) | | | |
| | Day 67 | Geological Processes and Rates - Questions<br>Ch 4: Worksheet 3 • Pages 51–52 • (TG) | | | |
| Week 5 | Day 68 | Geological Processes and Rates - Activities<br>Ch 4: Worksheet 3 • Page 52 • (TG) | | | |
| | Day 69 | Read Pages 48–53 • (GB) | | | |
| | Day 70 | Geological Processes and Rates - Words to Know<br>Ch 4: Worksheet 4 • Page 53 • (TG) | | | |
| | Day 71 | Geological Processes and Rates - Questions<br>Ch 4: Worksheet 4 • Pages 53–54 • (TG) | | | |
| | Day 72 | *Geology Book* Chapter 4 Study Day | | | |
| Week 6 | Day 73 | ***Geology Book* Chapter 4 Quiz 2** • Pages 113–114 • (TG) | | | |
| | Day 74 | Read Pages 54–57 • (GB) | | | |
| | Day 75 | Ways to Date the Entire Earth - Words to Know<br>Ch 5: Worksheet 1 • Page 55 • (TG) | | | |

| Date | Day | Assignment | Due Date | ✓ | Grade |
|------|-----|------------|----------|---|-------|
| Week 7 | Day 76 | Ways to Date the Entire Earth - Questions<br>Ch 5: Worksheet 1 • Pages 55–56 • (TG) | | | |
| | Day 77 | Ways to Date the Entire Earth - Activities<br>Ch 5: Worksheet 1 • Page 56 • (TG) | | | |
| | Day 78 | Read Pages 58–68 • (GB) | | | |
| | Day 79 | Great Geologic Events of the Past - Words to Know<br>Ch 6: Worksheet 1 • Page 57 • (TG) | | | |
| | Day 80 | Great Geologic Events of the Past - Questions<br>Ch 6: Worksheet 1 • Pages 57–58 • (TG) | | | |
| Week 8 | Day 81 | Great Geologic Events of the Past - Activities<br>Ch 6: Worksheet 1 • Page 58 • (TG) | | | |
| | Day 82 | *Geology Book* Chapters 5–6 Study Day | | | |
| | Day 83 | ***Geology Book* Chapters 5–6 Quiz 3** • Pages 115–116 • (TG) | | | |
| | Day 84 | Read Pages 69–72 • (GB) | | | |
| | Day 85 | Questions People Ask - Words to Know<br>Ch 7: Worksheet 1 • Page 59 • (TG) | | | |
| Week 9 | Day 86 | Questions People Ask - Questions<br>Ch 7: Worksheet 1 • Pages 59–60 • (TG) | | | |
| | Day 87 | Read Pages 73–75 • (GB) | | | |
| | Day 88 | ***Geology Book* Chapters 7–8 Quiz 4** • Page 117 • (TG) | | | |
| | Day 89 | *Geology Book* Chapters 1–8 Study Day | | | |
| | Day 90 | ***Geology Book* Test** • Pages 119–121 • (TG) | | | |
| | | Mid-Term Grade | | | |

# Second Semester Suggested Daily Schedule

| Date | Day | Assignment | Due Date | ✓ | Grade |
|------|-----|------------|----------|---|-------|
| | | Second Semester-Third Quarter — *The Cave Book* | | | |
| Week 1 | Day 91 | Read Pages 6–7 • *The Cave Book* • (CB) | | | |
| | Day 92 | Introduction - Words to Know, Short Answer, Discussion Questions<br>Intro: Worksheet 1 • Pages 63–64 • Teacher Guide • (TG) | | | |
| | Day 93 | Introduction - Activities<br>Intro: Worksheet 1 • Page 64 • (TG) | | | |
| | Day 94 | Read Pages 8–11 • (CB) | | | |
| | Day 95 | Read Pages 12–21 • (CB) | | | |
| Week 2 | Day 96 | Humans and Caves - Words to Know<br>Ch 1: Worksheet 1 • Page 65 • (TG) | | | |
| | Day 97 | Humans and Caves - Short Answer<br>Ch 1: Worksheet 1 • Pages 65–66 • (TG) | | | |
| | Day 98 | Humans and Caves - Discussion Questions<br>Ch 1: Worksheet 1 • Page 66 • (TG) | | | |
| | Day 99 | Humans and Caves - Activities<br>Ch 1: Worksheet 1 • Page 66 • (TG) | | | |
| | Day 100 | *Cave Book* Introduction–Chapter 1 Study Day | | | |
| Week 3 | Day 101 | ***Cave Book*** Introduction–Ch1 Quiz 1 • Pages 123–124 • (TG) | | | |
| | Day 102 | Read Pages 22–29 • (CB) | | | |
| | Day 103 | Caves and Mythology - Words to Know<br>Ch 2: Worksheet 1 • Page 67 • (TG) | | | |
| | Day 104 | Caves and Mythology - Short Answer<br>Ch 2: Worksheet 1 • Pages 67–68 • (TG) | | | |
| | Day 105 | Caves and Mythology - Discussion Questions<br>Ch 2: Worksheet 1 • Page 68 • (TG) | | | |
| Week 4 | Day 106 | Caves and Mythology - Activities<br>Ch 2: Worksheet 1 • Page 68 • (TG) | | | |
| | Day 107 | Read Pages 30–37 • (CB) | | | |
| | Day 108 | Caves and Karst - Words to Know<br>Ch 3: Worksheet 1 • Page 69 • (TG) | | | |
| | Day 109 | Caves and Karst - Short Answer<br>Ch 3: Worksheet 1 • Pages 69–70 • (TG) | | | |
| | Day 110 | Caves and Karst - Discussion Questions<br>Ch 3: Worksheet 1 • Page 70 • (TG) | | | |
| Week 5 | Day 111 | Caves and Karst - Activities<br>Ch 3: Worksheet 1 • Page 70 • (TG) | | | |
| | Day 112 | *Cave Book* Chapters 2–3 Study Day | | | |
| | Day 113 | ***Cave Book*** Chapters 2–3 Quiz 2 • Pages 125–126 • (TG) | | | |
| | Day 114 | Read Pages 38–47 • (CB) | | | |
| | Day 115 | Classifying Caves - Words to Know<br>Ch 4: Worksheet 1 • Page 71 • (TG) | | | |

| Date | Day | Assignment | Due Date | ✓ | Grade |
|------|-----|-----------|----------|---|-------|
| **Week 6** | Day 116 | Classifying Caves - Short Answer<br>Ch 4: Worksheet 1 • Pages 71–72 • (TG) | | | |
| | Day 117 | Classifying Caves - Discussion Questions<br>Ch 4: Worksheet 1 • Page 72 • (TG) | | | |
| | Day 118 | Classifying Caves - Activities<br>Ch 4: Worksheet 1 • Page 72 • (TG) | | | |
| | Day 119 | Read Pages 48–55 • (CB) | | | |
| | Day 120 | Exploring Caves - Words to Know, Short Answer<br>Ch 5: Worksheet 1 • Pages 73–74 • (TG) | | | |
| **Week 7** | Day 121 | Exploring Caves - Discussion Questions<br>Ch 5: Worksheet 1 • Page 74 • (TG) | | | |
| | Day 122 | Exploring Caves - Activities<br>Ch 5: Worksheet 1 • Page 74 • (TG) | | | |
| | Day 123 | *Cave Book* Chapters 4–5 Study Day | | | |
| | Day 124 | ***Cave Book* Chapters 4–5 Quiz 3** • Pages 127–128 • (TG) | | | |
| | Day 125 | Read Pages 56–67 • (CB) | | | |
| **Week 8** | Day 126 | Read Pages 68–72 • (CB) | | | |
| | Day 127 | Studying Caves - Words to Know<br>Ch 6: Worksheet 1 • Page 75 • (TG) | | | |
| | Day 128 | Studying Caves - Short Answer<br>Ch 6: Worksheet 1 • Pages 75–76 • (TG) | | | |
| | Day 129 | Studying Caves - Discussion Questions<br>Ch 6: Worksheet 1 • Page 76 • (TG) | | | |
| | Day 130 | Studying Caves - Activities<br>Ch 6: Worksheet 1 • Page 76 • (TG) | | | |
| **Week 9** | Day 131 | *Cave Book* Chapter 6 Study Day | | | |
| | Day 132 | ***Cave Book* Chapter 6 Quiz 4** • Pages 129–130 • (TG) | | | |
| | Day 133 | *Cave Book* Ch 1–6 Study Day | | | |
| | Day 134 | *Cave Book* Ch 1–6 Study Day | | | |
| | Day 135 | ***Cave Book* Ch 1–6 Test** • Pages 131–133 • (TG) | | | |
| | | Second Semester-Fourth Quarter — *The Fossil Book* | | | |
| **Week 1** | Day 136 | Read Pages 4–5 • *The Fossil Book* • (FB) | | | |
| | Day 137 | Introduction - Words to Know<br>Intro: Worksheet 1 • Page 79 • *Teacher Guide* • (TG) | | | |
| | Day 138 | Introduction - Questions<br>Intro: Worksheet 1 • Pages 79–80 • (TG) | | | |
| | Day 139 | Introduction - Activities<br>Intro: Worksheet 1 • Page 80 • (TG) | | | |
| | Day 140 | Read Pages 6–17 • (FB) | | | |

| Date | Day | Assignment | Due Date | ✓ | Grade |
|---|---|---|---|---|---|
| Week 2 | Day 141 | Fossils, Flooding, and Sedimentary Rock - Words to Know<br>Ch 1: Worksheet 1 • Page 81 • (TG) | | | |
| | Day 142 | Fossils, Flooding, and Sedimentary Rock - Questions<br>Ch 1: Worksheet 1 • Pages 81–82 • (TG) | | | |
| | Day 143 | Fossils, Flooding, and Sedimentary Rock - Activities<br>Ch 1: Worksheet 1 • Page 82 • (TG) | | | |
| | Day 144 | *Fossil Book* Introduction–Chapter 1 Study Day | | | |
| | Day 145 | ***Fossil Book* Introduction–Ch1 Quiz 1** • Pages 135–136 • (TG) | | | |
| Week 3 | Day 146 | Read Pages 18–25 • (FB) | | | |
| | Day 147 | Geologic Column Diagram - Words to Know<br>Ch 2: Worksheet 1 • Page 83 • (TG) | | | |
| | Day 148 | Geologic Column Diagram - Questions<br>Ch 2: Worksheet 1 • Pages 83–84 • (TG) | | | |
| | Day 149 | Geologic Column Diagram - Activities<br>Ch 2: Worksheet 1 • Page 84 • (TG) | | | |
| | Day 150 | Read Pages 26–29 • (FB) | | | |
| Week 4 | Day 151 | Read Pages 30–33 • (FB) | | | |
| | Day 152 | Flood Geology vs. Evolution - Words to Know<br>Ch 3: Worksheet 1 • Page 85 • (TG) | | | |
| | Day 153 | Flood Geology vs. Evolution - Questions<br>Ch 3: Worksheet 1 • Pages 85–86 • (TG) | | | |
| | Day 154 | Flood Geology vs. Evolution - Activities<br>Ch 3: Worksheet 1 • Page 86 • (TG) | | | |
| | Day 155 | *Fossil Book* Chapters 2–3 Study Day | | | |
| Week 5 | Day 156 | ***Fossil Book* Chapters 2–3 Quiz 2** • Pages 137–138 • (TG) | | | |
| | Day 157 | Read Pages 34–41 • (FB) | | | |
| | Day 158 | Read Pages 42–49 • (FB) | | | |
| | Day 159 | Kinds of Fossils I - Words to Know<br>Ch 4: Worksheet 1 • Page 87 • (TG) | | | |
| | Day 160 | Kinds of Fossils I - Questions<br>Ch 4: Worksheet 1 • Pages 87–90 • (TG) | | | |
| Week 6 | Day 161 | Kinds of Fossils I - Activities<br>Ch 4: Worksheet 1 • Page 90 • (TG) | | | |
| | Day 162 | *Fossil Book* Chapter 4 Study Day | | | |
| | Day 163 | ***Fossil Book* Chapter 4 Quiz 3** • Pages 139–140 • (TG) | | | |
| | Day 164 | Read Pages 50–55 • (FB) | | | |
| | Day 165 | Read Pages 56–64 • (FB) | | | |
| Week 7 | Day 166 | Read Pages 65–67 • (FB) | | | |
| | Day 167 | Kinds of Fossils II - Words to Know<br>Ch 5: Worksheet 1 • Page 91 • (TG) | | | |
| | Day 168 | Kinds of Fossils II - Questions<br>Ch 5: Worksheet 1 • Pages 91–92 • (TG) | | | |
| | Day 169 | Kinds of Fossils II - Activities<br>Ch 5: Worksheet 1 • Page 92 • (TG) | | | |
| | Day 170 | Read Pages 68–71 • (FB) | | | |

| Date | Day | Assignment | Due Date | ✓ | Grade |
|---|---|---|---|---|---|
| Week 8 | Day 171 | Conclusion - Questions<br>Conclusion: Worksheet 1 • Pages 93–94 • (TG) | | | |
| | Day 172 | Conclusion - Activities<br>Conclusion: Worksheet 1 • Page 94 • (TG) | | | |
| | Day 173 | Application - Read Pages 72–74 | | | |
| | Day 174 | Application - Read Pages 75–77 | | | |
| | Day 175 | Application - Questions<br>Application: Worksheet 1 • Pages 95–96 • (TG) | | | |
| Week 9 | Day 176 | Application - True/False<br>Application: Worksheet 1 • Page 96 • (TG) | | | |
| | Day 177 | *Fossil Book* Chapters 5–Conclusion Study Day | | | |
| | Day 178 | ***Fossil Book* Chapters 5–Concl Quiz 4** • Pages 141–142 • (TG) | | | |
| | Day 179 | *Fossil Book* Intro–Concl Study Day | | | |
| | Day 180 | ***Fossil Book* Intro–Concl Test** • Pages 143–146 • (TG) | | | |
| | | Final Grade | | | |

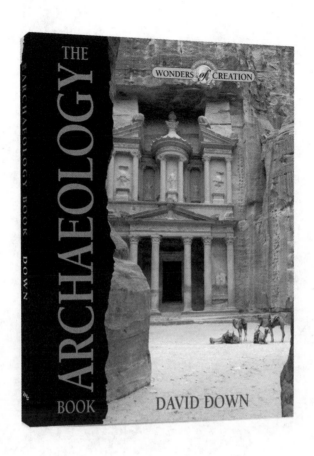

**Archaeology Worksheets**

**for Use with**

*The Archaeology Book*

# Words to Know

accession year

AD

archaeology

artifact

BC

carbon dating

ceramic

chronology

debris

EB

exile

exodus

hieroglyphs

LB

MB

millennium

non-accession year

pottery

synchronism

tell

## Questions

1. What does the word archaeology mean?

2. For what three reasons were cities built on hills?

3. When did people first start using coins?

4. Why are inscriptions found on ancient pottery valuable to archaeologists?

5. What are the four main periods of archaeological time?

## Activities

1. See if you can find a small piece of damp clay, or plasticine, and with the end of a screwdriver impress your name on it. This would then look like a seal impression.

2. Take some everyday items and set up an archaeological treasure hunt. Have an adult bury the items in shallow holes, covering them with a thin layer of soil. Carefully go about digging them up and classifying your treasures in a journal.

## Words to Know

Asiatic

baulk

dowry

drachma

dynasty

mastabas

Nubia

Pharaoh

## Questions

1. What is the Egyptian name for Egypt?

2. Who was the first Egyptian to build a pyramid?

3. Who built the biggest pyramid in Egypt?

4. What was the name of the Egyptian god of the Nile River?

5. What did the Egyptians in dynasty 12 mix with their bricks to hold them together?

## Activities

1. See if you can find a small cardboard or plastic box. Make some mud out of earth and a little water, mix some dry grass with it, and put it into the box. When it is fairly dry, turn the box upside down and lift it off the brick you have made. Let it dry.

2. Develop a chart with your family history or dynasty. Try to trace the ancestry of one parent or both, depending on the information you have available. List these as names on a graph or draw an actual tree with the branches representing family members.

## Words to Know

amphitheater

Anatolia

bathhouse

inscription

## Questions

1. Which was the strongest nation in the Middle East 3,000 years ago?

2. Which two nations did the Syrians think had come to attack them?

3. Who were the Hittites descended from?

4. How often were the Hittites mentioned in the King James Version of the Bible?

5. Who wrote the book *The Empire of the Hittites*?

## Activities

1. Draw a rough map of Turkey and write in the names Constantinople and Boghazkale where you think they should be.

2. Read the book of Esther. Write three discussion questions about the story and find an evening this week to discuss the story with your family.

## Words to Know

centurion

Chaldees

nomad

papyrus

## Questions

1. In the Bible, how many references are there to Ur of the Chaldees?

2. Who was the main excavator of Ur of the Chaldees?

3. Why did Woolley not excavate the cemetery as soon as he found it?

4. What was the name of the people who occupied ancient Ur?

5. What did Woolley find in the Death Pits of Ur?

## Activities

1. Research the ancient Hittite civilization online or at your local library. How many resources can you find for this people once thought to be a myth?

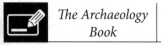
## Words to Know

bulla

Medes

scarab

seal

## Questions

1. Who discovered Nineveh?

2. What was the name of the ruins where Layard first started digging?

3. What was the name of the king of Israel that was mentioned on the black pillar Layard found in Nimrud?

4.  What was the name of the king of Israel when Sennacherib besieged Jerusalem?

5.  How many cities did Sennacherib claim he conquered?

## Activities

1.  Get some plasticine or soft clay and press it down flat. Then get a pencil or small twig the width of a pencil. Cut one end to form a triangle and press this end down horizontally and vertically on the clay. You will thus be making cuneiform impressions.

2.  Sometimes archaeologists must learn a language to help them understand a culture better, just as Layard studied the Persian language. Choose a language to study briefly and obtain several books from the library to help you learn some basic words and phrases.

## Words to Know

Armenians

cuneiform

strata

syncline

## Questions

1. What was the name of the cuneiform record that told a story similar to the Bible record of Noah and the Flood?

2. Which Assyrian king compiled a library of tablets in Nineveh?

3. What did the Babel builders stick their bricks together with?

4. Which king made Babylon a city of gold?

5. Which Bible prophet predicted that Babylon would become uninhabited?

## Activities

1. Log on to the Internet with a parent's permission and search for "Ishtar Gate Berlin Museum." This will bring up a picture of the gate from Babylon that Professor Koldewey sent back to Berlin.

2. Research the Bible account of the Flood and compare this account with other accounts from around the world. A good resource from a Christian perspective is *Flood Legends* by Charles Martin, or *The Flood of Noah*, available from Master Books.

## Words to Know

Persia

rhyton

## Questions

1. Who was the king who first carved out the Medo-Persian Empire?

2. In what year did he conquer Babylon?

3. Which Persian king left an inscription on the rock face of the Zagros Mountain near Bisitun?

4. What was the name of the great Persian city that Darius built?

5. What was the name of the official who tried to destroy all the Jews in Persia?

## Activities

1. Read the book of Esther in the Bible and count how many times the word God is used. You may be surprised.

2. Study the celebration of Purim that is still celebrated today. Observe how the traditions and even the games relate back to Queen Esther.

## Words to Know

Bedouin

cistern

Edom

Edomites

Nabataeans

siq

theater

wadi

## Questions

1. In what year did Burckhardt discover Petra?

2. Whose descendants occupied Petra?

3. What were his descendants called?

4. Which Bible prophet wrote a book about Petra?

5. Which Roman emperor had a road made through Petra?

## Activities

1. Pottery in Petra was very thin. Get some plasticine or clay and make a small teacup without a handle. See how thin you can make it.

2. Set up a tent in your yard and talk about what it would be like to live life as a Bedouin, wandering from place to place. Consider staying overnight in the tent, weather permitting.

## Words to Know

Baal

causeway

Yehovah

## Questions

1. What were the four main cities of ancient Phoenicia?

2. What kind of trees did Solomon pay the Phoenician King Hiram for?

3. Whose tomb did Pierre Montet find?

4. Which Bible prophet challenged the prophets of Baal?

5. Which Bible prophet predicted that ancient Tyre would never be found?

## Activities

1. Find a map of the Mediterranean Sea and try to work out how far it is from Phoenicia (modern Lebanon) to Spain. That is how far Phoenician ships sailed.

2. Make a relief of your hand by pressing your palm and fingers into a flat piece of clay or by pressing foil over your hand to make a metallic-looking imprint. See how much detail you can add once the initial impression is made.

## Words to Know

scroll

vellum

## Questions

1. In what year was the first Dead Sea Scroll found?

2. How many letters were in the Hebrew alphabet?

3. What were most of the Dead Sea Scrolls written on?

4. Psalm 119 is a form of Hebrew poetry called what?

5. What was the name of the settlement near the cave where the Dead Sea Scrolls were found?

## Activities

1. Find a King James version of the Bible and look at Psalm 119. At the beginning of every eighth verse you will find a letter of the Hebrew alphabet. Try to write out the 22 letters of the alphabet.

2. Take several pieces of white or tan cardstock. Write or paint a favorite Bible passage across it. When dry, roll up the "scroll" and tie it off with ribbon or string.

## Words to Know

annunciation

Calvary

Golgotha

grotto

Messiah

ossuary

Passover

## Questions

1. Which Roman emperor adopted Christianity as the state religion?

2. Jesus' name in Hebrew was Yeshua. What does it mean?

3. In what river was Jesus baptized?

4. In which city did Jesus enter a synagogue and cast out a demon?

5. What does the word Calvary mean?

## Activities

1. At the back of most Bibles are maps of Palestine. Try to calculate how far it was from Jerusalem to Galilee. Jesus walked this distance many times.

2. Using poster board or cardstock, create a map of Israel. You might consider paints or markers to color the rivers, lakes, and land. Glue on small blocks or other objects to represent towns and cities.

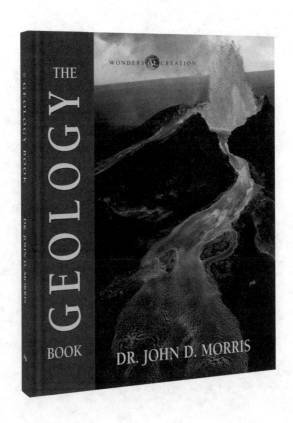

**Geology Worksheets**

**for Use with**

*The Geology Book*

**Scripture:** Genesis 1:1–31; Genesis 3:17–21; Romans 6:23; Romans 8:22

## Words to Know

Principle of uniformity

Principle of catastrophe

Asthenosphere

Plate

## Questions

1. Operational science is the science that deals with repeatable, observable experiments in the present. Origins science deals with reconstructing events that have happened in the past. What is the key difference between "origins" and "operational" science?

2. There are two ways of thinking about the unobserved past. What are they?

3. Where is the true history of the earth found?

4. In what order did God create the heavens and earth? (e.g., describe what He created on Day 1, Day 2, etc.) See Genesis 1.

5. Write a short paragraph answering the question, "What is sin?"

6. What are the main "zones" into which the earth is divided?

7. What is the earth's crust composed of?

8. What is the purpose of the earth's atmosphere?

## Activities

Review the text on pages 4–9 again. Two views of earth history are compared (uniformity and catastrophe). Make a chart of the comparisons. See if you can find three to five examples to include in your comparison.

**Scripture:** Genesis 1:1; Obadiah 1:3

## Words to Know

Igneous rock

Sedimentary rock

Metamorphic rock

Ripple marks

Crossbed

Concretions

Metamorphism

## Questions

1. This chapter lists three categories of rock, with each category containing a discussion on several types of rock. Draw an expanded version of the table on the next page.

   a. In the first column, list each type of rock mentioned in this chapter.

   b. In the second column, list the category under which the rock is found.

   c. In the third column, describe the composition of each rock type.

   d. In the fourth column, describe how the rock is formed.

   e. In the fifth column, make a list of where the rock is found today.

   f. Watch out for types within types! (We've done the first one for you!)

| Type | Category | Composition | Formation | Found |
|---|---|---|---|---|
| Granite | Igneous | Quartz and feldspar with mica and hornblende | Formed when molten rock is cooled | Mountains Upper mantle |
| | | | | |
| | | | | |
| | | | | |
| | | | | |
| | | | | |
| | | | | |
| | | | | |

## Activities

Start collecting stones/small rocks from around your area (or other areas to which you travel). Try to classify the type of rock you have found. Can you find samples of each rock you described in the above table?

**Note:** If you go to a national/state/local park, please ask permission to remove the stones/rocks you are collecting. Do not remove any rocks or stones from someone's garden without permission.

**Scripture:** Genesis 8:4; Psalm 121

## Words to Know

Plain

Sediment

Plateau

Mountain

Canyon

## Questions

1. Why are low-lying plains considered good for farmland?

2. Three types of plateaus are mentioned. Write a short description and give an example of each type.

3. Four types of mountains are mentioned. Write a short description and give at least one example of each type.

4. What process causes the formation of mesas and buttes?

## Activities

This "experiment" will take a few weeks. Erect a mound of dirt in your backyard (pile of dirt should be at least three feet high). Visit the mound each day and record the following information: height of mound, width of mound. You will notice that the mound will get shorter and the base wider. Determine what could have caused the difference in height and width. Was it the wind? Was it rain? Was it a dry spell? Have fun with this.

**Scripture:** Genesis 8

## Words to Know

Erosion

Deposition

Turbidite

## Questions

1. List the five primary causes of normal erosion.

2. List and describe the three rapid erosive processes explained in this chapter.

3. How is a turbidite formed?

4. List two types of events that could cause deposition to happen quickly and on a large scale.

## Activities

Observe and record the types of erosion you find near where you live. If possible, take pictures to document your findings.

**Scripture:** Genesis 7, 8

## Words to Know

Compaction

Cementation

Fossils

Petrification

Gastrolith

Coprolite

## Questions

1. Sand is made primarily of what mineral?

2. Write a paragraph describing how sedimentary rocks are formed.

3. What is the main condition required for a fossil to form?

4. There are many different types of fossils. Name at least four.

5. In your own words, write a description of how dinosaur fossils were formed.

## Activities

**Note:** This week's activities reinforce the fact that fossils and petrified wood can be made over a short period of time.

- Make a "fossil" using plaster of paris. Mix the plaster of paris and pour it onto a paper plate. Gather different objects (leaves, toy dinosaurs, etc.). Press each into the plaster of paris and then lift the object off. Include a print of your hand or foot. Let dry. Paint when dry, if desired.

- This is week 3 of the experiment started in Chapter 3. Has your mound decreased any? What caused this decrease?

**Scripture:** Genesis 7, 8

## Words to Know

Volcanism

Fumaroles

Geyser

Fault

Earthquake

## Questions

1. At least two ways that volcanoes erupt are discussed in the text. What are they?

2. When forces in the earth's crust build up to a breaking point along a fault, the sections move in one of three ways. What are they? Write a short description of each type.

3. What factors influence whether a rock will bend or break?

4. Write two paragraphs explaining the arguments given in the book for and against continental separation.

## Activities

1. You might want to do a more in-depth study of volcanoes. Research some of the more famous volcanoes that have erupted in the past. When did they erupt? How devastating to the surrounding areas were they? What types of devastation did they cause? How often do these volcanoes erupt? How long do the eruptions last?

2. This is week 4 of this experiment started in Chapter 3. Has your mound decreased any? What caused this decrease?

**Scripture:** Psalm 18

## Words to Know

Atom

Isotope

Radioisotope dating

Carbon dating

## Questions

1. In your own words, describe the different ways metamorphic rocks are thought to form.

2. What are unstable atoms called? What is the most well-known radioactive atom?

3. How are the unstable qualities of uranium useful to mankind?

4. What is the difference between a parent isotope and a daughter isotope?

5. Explain the process of carbon dating in determining when a plant died.

**Scripture:** Psalm 18

## Word to Know

Magnetic field

Uplift

## Questions

1.  List some of the methods currently being used to determine the age of the earth.

2.  What must we consider when evaluating conclusions obtained from these methods?

3. What conclusion can be drawn about the age of the earth from the various dating methods discussed in this chapter?

    **Note:** Be sure to read the picture captions!

## Activities

This is a simple experiment to study wind erosion. Stick a piece of two-sided tape on one side of several different stirring paddles (you can get these from a paint store) and place them in the ground in various places around your yard. Make sure they don't all face in one direction. Have one face north, another south, etc. At regular intervals, check the amount of dust or soil sticking to the tape. Depending on the amount of wind and the direction from which it blows, you will see that some paddles collect more dust than other paddles. More soil will stick to the paddles where wind erosion is taking place. What is in the way that is preventing dirt from sticking to those paddles where little is collected?

**Scripture:** Genesis 1–11; Romans 6:23

## Words to Know

Second law of science

Fountains of the deep

Glacier

Polar ice cap

## Questions

1. What four events have had the greatest impact in shaping the earth's geology?

2. How did the creation event affect the earth's geology?

3. What role has the Fall played in shaping today's earth?

4. What was the cause of the global Flood? What were some of the geological results of the Flood?

5. Why do we find ocean fossils near the top of Mt. Everest?

6. Draw a map of North America. Outline the extent of the ice covering at the peak of the Ice Age. (Hint: be familiar with the map on page 67.)

## Activities

Write a two-page comparative essay on the cause of the Ice Age. Compare the views presented in *The Geology Book* with secular views on this subject. Your essay should include an introduction, your thesis statement, an explanation of each viewpoint, and your conclusion. Independent research of Christian and secular viewpoints will be required to complete this activity.

Answers in Genesis (www.answersingenesis.org), the Creation Research Society (www.creationresearch.org), and the Institute for Creation Research (www.icr.org) are recommended resources.

## Word to Know

Volcanism

Escarpment

## Questions

**Note:** This week's questions are the same as the questions in the text. It is important that the main concept of each question is expressed in the answer. The answers should be worded as the student understands and can apply them rather than simple memorization.

1. How was Grand Canyon formed?

2. What causes the geysers in Yellowstone Park?

3. How did Niagara Falls form?

4. Why are the Appalachian and Rocky Mountains so different?

5.  How long does it take to form petrified wood?

6.  How are stalactites and stalagmites formed?

7.  How is coal formed?

8.  How is natural gas formed?

9.  How is oil formed?

10. Are dinosaur fossils the most abundant type of fossil?

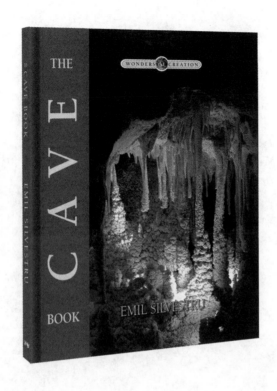

**Cave Worksheets**

**for Use with**

*The Cave Book*

## Words to Know

karst

karst aquifiers

## Short Answer

1. What is the probable reason some of our ancestors may have entered the "underland" of cave systems?

2. The strange event near the _____ not only split the once unified population, but also scattered those with different skills and abilities.

3. How much of the world's drinking water comes from limestone (karst) terrains?

4. How much is it estimated to be by 2025?

## Discussion Questions

1. What role did caves play for our ancestors?

2. How did the events surrounding the Tower of Babel affect the ancient groups of people who dispersed from that area?

## Activities

1. Read through the account of the Tower of Babel (Genesis 11:1–9) and discuss issues that would have affected the various people groups when this event occurred.

2. Use *Adams' Chart of History* to trace the lineage of various people groups from the time of the Tower of Babel to the early 1900s.

## Words to Know

acoustics

Acheulean industry

bas-reliefs

cave paintings

engravings

Kyr

Myr

speleothems

Neanderthals

## Short Answer

1. Were there caves present before the Flood?

2. When does the Bible mention caves for the first time?

3. How many times is the word "cave" mentioned in the Bible?

4. Name three large animals that lived in caves prior to their extinction.

5. In what country is Longgupo Cave, which is believed to host the oldest stone artifacts?

6. What is the Twin River Cave in Zambia known for: (a) oldest human remains, (b) oldest burial site, or (c) art associated with burial rituals?

7. What are the three kinds of cave art that have been found?

8. Were Neanderthals a different species than us?

## Discussion Questions

1. Why did humans move to caves after having lived in cities?

2. Why did humans worship inside caves?

3. Where is the largest number of cave paintings located? Why?

4. What is the main criterion to separate various human cultures of the past?

5. Who were the Neanderthals?

6. What did the human remains from Shanidar Cave reveal about Neanderthal social life?

## Activities

1. Do a keyword search for the word "cave(s)" in the Bible using a Bible concordance or online research (with a parent's permission. Examine the various reasons people used caves during the biblical period.
2. Use a plastic knife to carve a bar of soap into a stone axe head. Examine the different uses for an axe in a book or online research (with a parent's permission).
3. With a large sheet of paper and paints (or chalk), recreate the look of cave paintings. You might consider using a dark room or garage with candles (under parental supervision) to sense the dark cave atmosphere.

## Words to Know

arthropods

bidirectional air circulation

cul-de-sac

echolocation

troglobites

troglophiles

trogloxenes

unidirectional air circulation

## Short Answer

1. Name at least three civilizations that have caves present in their mythology.

2. Which is the largest troglobite alive today?

3. Are bats: (a) trogloxenes, (b) troglophiles, or (c) troglobites?

4. What is the Movile Cave in Romania famous for?

5. What is the normal humidity inside most caves?

## Discussion Questions

1. What happens when female bats give birth?

2. What is the longest period of cave habitation in modern times? Why did those humans choose to live inside a cave?

3. How can ice accumulate inside caves in a temperate climate?

4. What are the health benefits of caves?

## Activities

1. The Dead Sea Scrolls contained fragments from the Old Testament books of Genesis, Exodus, Leviticus, Numbers, Deuteronomy, 1 and 2 Samuel, Psalms, Job, Isaiah, Daniel, Jeremiah, Ezekiel, and parts of the Minor Prophets. Pick a passage and write it out on a piece of paper. Roll it up to reconstruct the look of a scroll fragment that would have been discovered in the Qumran cave system.

2. Build a cave model with clay. Try to include forms that represent bidirectional airflow, cold air traps, and unidirectional airflow passages.

3. Do further research on cave creatures (fauna) and list the various types of creatures one might expect to find in a cave system.

| | *The Cave Book* | Caves and Karst Pages 30–37 | Day 108, 109, 110, & 111 | Chapter 3 Worksheet 1 | Name |
| --- | --- | --- | --- | --- | --- |

## Words to Know

cave

emergences

endogenetic

exogenetic

karsted

orthokarst

parakarst

pseudokarst

resurgences

sinkholes

## Short Answer

1. What are karstic rocks?


2. What percentage of the dry, ice-free landmass is covered by karstic rocks?


3. Besides limestone, what other sedimentary rocks host many caves?


4. What is a resurgence in karst terrains?


5. What is an emergence in karst terrains?

6. What is the highest average flow of a karst river? Compare that to the daily water consumption of New York.


7. What is a rhythmic spring? Name one.


## Discussion Questions

1. What are characteristics of caves found in igneous rocks?


2. Where does the name "karst" come from?


3. Name a few specific forms of the karst relief.


4. Where in South America has a surprising parakarst been discovered? What rock is it developed in?


5. What is unusual for the karst terrain in the Guadalupe Mountains in New Mexico?


## Activities

Complete online research (with a parent's permission) on cave systems around the world. Compile information on at least one system per continent, print off photos of each one, and use a world map to pinpoint their locations.

## Words to Know

active caves

compoundrelict caves

denudation rate

detrital formations

dripping speleothems

phreatic caves

relict caves

vadose caves

## Short Answer

1. What is an active cave?

2. How many types of active caves are there? List them.

3. What is a shield?

4. How many kinds of eccentric speleothems are there? What is the criterion to classify them?

5. What are cave rafts?

6. Name at least three non-calcite speleothems.

## Discussion Questions

1. Outline in very simple terms how speleothems are dated.

2. What is the most obvious and logical argument against a very old age of speleothems?

## Activities

1. Build a three-dimensional cave model using a cardboard box, such as a shoebox. Draw small pictures of the various cave formations (including stalagmites, stalactites, columns, flowstones, cave coral, etc.), then glue or tape the various drawings into the cardboard shell, labeling each structure.

2. Find old or broken items (with a parent's permission) and have someone bury them in your backyard. Set up a specific perimeter with strings staked in the ground (as they would in an archeological excavation site). Carefully dig up the "artifacts" and catalog where they were found and what use they may have served.

## Words to Know

desiccation cracks

master joints

scallops

## Short Answer

1. Who was the first to build and use a diving device?

2. Who was the one who co-invented the Aqualung and was the first to SCUBA dive inside a cave?

3. What family spent the most time in a cave during World War II?

4. What is the name of the cave system this family used?

## Discussion Questions

1. Describe the gear needed for cave exploration.

2. Discuss some of the concerns, problems, and challenges faced by modern cave explorers as they study a cave.

3. What concerns are there for those trying to take photographs in a cave?

## Activities

Write a two-page paper describing what you think life would have been like for the Stermer family. Consider the physical, emotional, and spiritual aspects of hiding in a cave for two years. What can we learn from their story?

## Words to Know

cenote

concavities

convexities

diagenesis

Xibalba

## Short Answer

. What is the only measurement of the karsting processes generally accepted?

. What are the two basic types of waters that reach karstic rocks?

. Extremely valuable information is yielded by continuous measurements of the discharge or flow rates (_____) and chemical composition (_____) of waters emerging through karst springs.

. How are karst aquifers viewed today?

5. Within what amount of time after the Flood did the Ice Age set in?

## Discussion Questions

1. Describe the various areas of science covered in the study of caves.

2. Why do karstologists never recommend any significant mining activities below the karst water table?

3. Discuss why karstlands are one of the most sensitive types of environments.

## Activities

1. After completing a study of caves, consider taking a tour of a local cave system. Use this opportunity to discuss the various topics you have learned about karsts, including formation, history, and the life contained inside.

2. Create a diagram of the three stages of the creationary model for cave formation. See page 65 of *The Cave Book*.

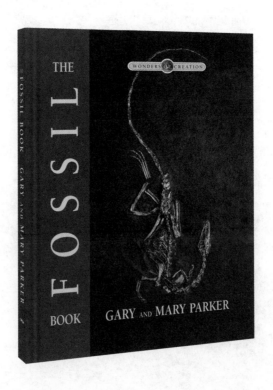

**Fossil Worksheets**

**for Use with**

*The Fossil Book*

## Words to Know

Evolution

Paleontologist

## Questions

1. When did most of the branches of modern science begin?

2. True/False (If false, explain your answer.) Most of the scientists who began the modern sciences firmly believed in a biblical history.

3. Who were the two men given credit for popularizing the modern teaching of evolution?

4.  What does TCSD stand for?

5.  List and describe the four C's of biblical history discussed in this book.

## Activities

Who was Charles Lyell? Research his life and his impact on long-age ideas. Write a short paper summarizing your findings.

## Words to Know

Archeology

Artifacts

Fossil

Geology

Paleontology

Permineralized fossils

Polystrates

Pseudofossils

Sediments

Trace fossils

## Questions

1. What types of sedimentary rocks are fossils normally found in?

2. List two agents that erode and deposit sediments.
    a. _____
    b. _____

3. Which agent is more powerful?

4. What two elements must exist in the right amounts for sediments to turn into rocks?
    a. _____
    b. _____

5. True/False (If false, explain your answer.) Time is a vital part of rock or fossil formation.

6.  What are the two most common rock cements?

    a. _____

    b. _____

7.  Give three examples where calcium carbonate can be found.

    a. _____    c. _____

    b. _____

8.  Give one example where silica can be found.

9.  What type of event would provide the right conditions to form fossils?

10. Briefly describe how a fossil can form.

11. Why must fossilization begin quickly?

12. What is the most common type of fossil?

13. What is the difference between permineralized wood and petrified wood?

14. Why is coal considered to be a fossil fuel?

15. Describe how coal forms (based on the research of Dr. Steve Austin).

16. What catastrophic event in May of 1980 supports Dr. Austin's theory?

17. How do polystrate fossils imply rapid burial?

## Activities

1.  Write a short research paper describing how coal is formed from a Bible-based perspective.
2.  Take a field trip to your local natural history museum. How many of the different types of fossils described in this chapter are on display there? Take along a sketch pad and pencil and draw a representative from the different fossil types (permineralized, mold, cast, carbon films, preserved soft parts, amber, trace, etc.). Label each drawing with the name of the fossil, the type of fossil, and where it was found.
3.  Find out if you live near a coal mine that offers tours and plan a visit.

## Words to Know

Geologic column

Index fossil

Living fossils

Trilobite

## Questions

1. In what type of rock are most fossils found? Where can we see these layers of rock?

2. What do evolutionists claim the geologic column represents?

3. According to Flood geologists, what does the geologic column show?

4. How many major geologic systems have been named? How many "super systems"?

5. Explain why fossils of sea creatures are found throughout the geologic column while animals and land plants tend to be found higher in the column.

6. What is the difference between how Flood geologists and evolutionists use the words "first" and "last"?

7. Who said that fossils are "perhaps the most obvious and serious objection to the theory of evolution"? Why is this significant?

## Activities

Begin researching what types of rock layers and fossils are prevalent in your area. Do you have pre-Flood, Flood, or post-Flood rocks?

## Words to Know

Cambrian explosion

Cavitations

Paraconformities

Stromatolites

## Questions

1. What is the lowest system in which fossils are found?

2. How have evolutionists tried to deal with the complex life seen in the fossils found in Cambrian layers?

3. True/False (If false, explain your answer.) Fossils found in pre-Cambrian rock are non-complex life forms.

4. About how much of a volcano's eruption emission is water vapor?

5.  What tilted the basement rocks of the earth's surface?

6.  How do the crystalline basement rocks found in Grand Canyon give testimony against millions of years of erosion?

7.  Describe the "breached dam" concept for the formation of Grand Canyon.

8.  What event supported the Flood geologists' interpretation of Grand Canyon's formation?

## Activities

Find out more about Flood geology by visiting www.AnswersInGenesis.org/go/geology or www. AnswersInGenesis.org/go/fossils. Or watch *Biblical Geology: Properly Understanding the Rocks* or read *Grand Canyon: A Different View.*

## Words to Know

Arthropod

Cephalopods

Diatoms

Echinoderms

Gastropods

Invertebrate

Malacology

Mollusks

Nautiloids

Palynology

Protozoan

Spicules

## Questions

1. Over 95% of all fossils found are _____.

2. True/False (If false, explain your answer.) Sometimes snail shell lids and the rest of the shell are found in two totally different layers of sediment.

3. How do fossil clams testify to a rapid burial?

4. What is perhaps the best mollusk evidence of creation and Flood geology?

5. Why do some claim the nautiloid is proof of evolution?

6. How does the evolutionary claim in question 5 fall short?

7. What is the largest coral reef in the world?

8. True/False (If false, explain your answer.) Present-day coral reefs could not have grown to their present size in the few thousand years since the Flood.

9. Which layer of the geologic column is referred to by evolutionists as the Age of Crinoids and Flood geologists as the Zone of Crinoids?

10. How do the crinoid fossils found in central United States testify to the Flood?

11. Insect fossils are rare since their hard outer layer decomposes easily after death. However, insect fossils with delicate details have been found. How is this possible?

12. What are the first animals fossilized in abundance?

13. How does the complexity of these first fossils disprove evolution?

14. How did fossils of sea creatures end up on the top of mountains?

## Activities

Make your own fossil.

**Materials needed:** Plaster of paris or playdough, paper plates or aluminum pie plates, objects to press into medium (leaves, dinosaur toys, etc.)

1. Pour the plaster of paris into aluminum pie plates or paper plates (or pass out jars of playdough and a paper plate) — one plate or jar per child. Have children press objects (or hands) into the plaster or playdough and lift them off, leaving the imprint behind. Allow to dry. Paint if desired.

2. Bury your fossil and go on an excavation. Use the proper tools and techniques (found in the "Application" section of this book) and practice extracting your fossil.

## Words to Know

Evolutionary series

Metamorphosis

Splint bones

Vertebrates

## Questions

1. What are the five groups of vertebrates?

   a. _____   d. _____

   b. _____   e. _____

   c. _____

2. Which of the following has the most DNA per cell of any other animal group?

   a. amphibians
   b. birds
   c. fish
   d. reptiles
   e. mammals

3. Coelacanths were once thought to be a transitional form between fish and amphibians. How was this claim disproved?

4. Explain what the Bible teaches about animals that eat meat today.

5. Discuss how dinosaurs can be explained by the biblical account of creation and the Flood.

6. What were the dimensions of the Ark?

7. What is likely about the dinosaurs on Noah's Ark during the Flood?

8. Discuss two biblical explanations for the extinction of dinosaurs.

    a.

    b.

9. Why should *Archaeopteryx* not be considered a missing link?

10. Explain why the alleged sequence of horse hooves does not prove evolution.

11. Choose your favorite fossil and explain how it can be a "missionary fossil."

## Activities

1. Find out more information on the Ice Age by reading *The Weather Book* and *Life in the Great Ice Age*.
2. Find out more about the so-called "transitional forms" by visiting www.AnswersInGenesis.org/go/fossils. Choose one alleged transitional series (e.g., horse, whale, apeman, dinosaur-to-bird) to research in detail. Write a short paper detailing your findings.

## Questions

1. Explain how the fossil record supports the biblical account of creation.

2. Explain how the fossil record supports the biblical account of the corruption of the earth.

3. Explain how the fossil record supports the biblical account of a global Flood.

4. Explain how the fossil record supports the biblical account of God's mercy on His creation.

## Activities

1. Using the information you've learned so far and in the "Application" section of this book, plan a trip to a nearby cliff, cut, creek, or quarry to hunt for fossils. If you don't live in an area that is fossil-rich, consider spending your next family vacation in such a place. Make sure you have the proper permits, if applicable!

2. Creation Studies Institute offers exciting fossil-finding expeditions. For more information on their tours, visit www.CreationStudies.org.

3. Read the newspaper for articles about newly-discovered fossils. Clip out each article, place them in a journal, and then critique each one from a biblical perspective.

## Questions

1. What are the best types of rocks in which to find fossils?

2. Where are the four major places to find fossils?

3. Since fossils are so old, does that mean you can always look for them wherever you want? Explain your answer of yes or no.

4. What about fossils that you find on your own property — can you collect those?

5. Why are quarries such great places to find fossils?

6. Why is it important that you get special permission and take great care when searching quarries for fossils? (Caution: Always have experienced, adult supervision — and never go alone!)

7. List four things you need to take fossil hunting to protect any fossils that you find:
   a.
   b.
   c.
   d.

8. What is wet screening, and what type of environment is best to use this technique to find fossils?

9. What is the material that is attached to and surrounds a fossil called?

10. Why is it important not to chip away too closely to a fossil?

11. Label the following illustration showing the major invertebrate groups and ferns:

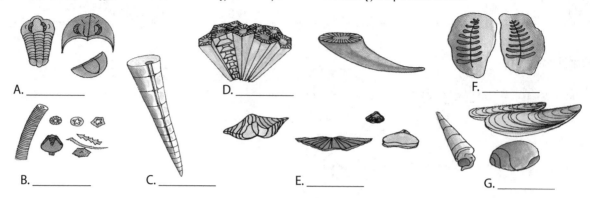

A. _____

B. _____

C. _____

D. _____

E. _____

F. _____

G. _____

12. Why are photos more important in identifying your fossil than knowing long scientific names?

13. Do you need special display cases or chemicals to prep fossils for storage?

14. Match the fossil with the biblical application by drawing lines:

   a.  Nautilus                   Corruption

   b.  Trilobites or fossils with bite marks   Creation

   c.  Living Fossils             Catastrophe

   d.  Closed fossil clams        Christ

## Bonus Question

Look up the phrase "living fossil." How do living fossils make clear that an evolutionary timescale is not possible?

## True/False Questions

1. T/F   Fossils are found in rocks that have been very hot.

2. T/F   Sedimentary rocks are not found in layers.

3. T/F   Igneous and metamorphic rocks are good places for fossils.

4. T/F   State parks do not allow individuals to collect fossils.

5. T/F   Private landowners can give you permission to search for fossils along road cuts.

6. T/F   Any type of tool will work in any location when it comes to excavating fossils.

7. T/F   Sometimes you may need to use the plaster jacketing technique to protect fossils.

8. T/F   Large fossils in plaster jackets do not need splints to stabilize them.

9. T/F   Always wear goggles when chipping at the matrix surrounding a fossil.

10. T/F   Fossil teeth are rarely preserved.

11. T/F   The edges of fossil teeth can make identification difficult.

12. T/F   Mastodon and mammoth teeth look the same.

13. T/F   Fragments of fossil bone are very difficult to identify.

14. T/F   Fossils are a great example of evidence for Noah's Flood.

15. T/F   Some of the earliest fossils are the most simple in structure.

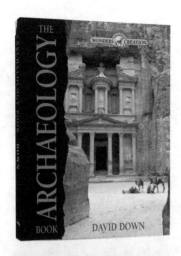
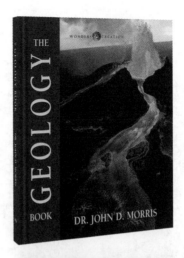
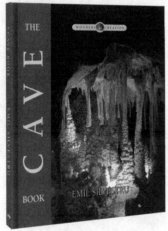

**Quizzes & Tests**

**for Use with**

*General Science 2: Survey of Geology & Archaeology*

| **Q** | *The Archaeology Book*<br>Concepts & Comprehension | Quiz 1 | Scope:<br>Chapters 1–2 | Total score:<br>____of 100 | Name |

## Define: (5 Points Each Answer)

1. accession year: _____

2. AD: _____

3. BC: _____

4. carbon dating: _____

5. EB: _____

6. LB: _____

7. MB: _____

8. baulk: _____

9. synchronism: _____

10. mastabas: _____

## Multiple Answer Questions: (2 Points Each Blank)

11. What are the four main periods of archaeological time?

    a. _____     c. _____

    b. _____     d. _____

12. For what three reasons were cities built on hills?

    a. _____

    b. _____

    c. _____

## Short Answer Questions: (4 Points Each Question)

13. What does the word archaeology mean? _____

_____

14. When did people first start using coins? _____

_____

15. What was the name of the Egyptian god of the Nile River? _____

_____

16. What is the Egyptian name for Egypt? _____

_____

17. Who was the first Egyptian to build a pyramid? _____

18. Who built the biggest pyramid in Egypt? _____

_____

## Applied Learning Activity: (12 Points Total; 1 Point Each Answer)

19. Identify the Pyramids, Temples, Tombs, and unique features on Giza Map:

Pyramid of Khufu

Valley Temple of Khufu

Pyramid of Menkaure

Valley Temple of Menkaure

Pyramid of Kahfre

Valley Temple of Kahfre

The Sphinx

Temple of the Sphinx

Pyramids of Queens

Queen's Tombs

Eastern Cemetery

Mortuary Temple

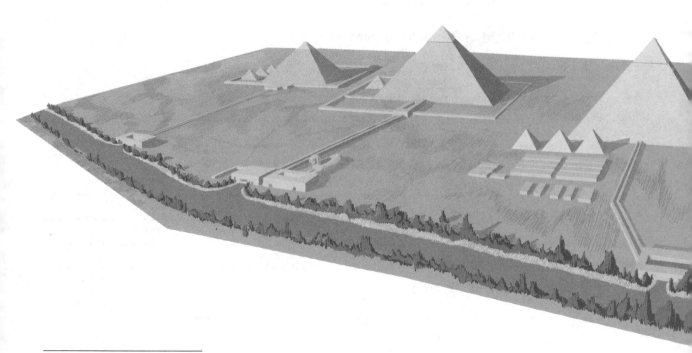

| Q | The Archaeology Book<br>Concepts & Comprehension | Quiz 2 | Scope:<br>Chapters 3–5 | Total score:<br>____of 100 | Name |
|---|---|---|---|---|---|

## Define: (5 Points Each Answer)

1. amphitheater: _____

2. Anatolia: _____

3. centurion:_____

4. Chaldees: _____

5. bulla: _____

6. scarab: _____

## Multiple Answer Questions: (4 Points Each Blank)

7. What two nations did the Syrians think had come to attack them?

    a. _____    b. _____

## Short Answer Questions: (4 Points Each Question)

8. Which was the strongest nation in the Middle East 3,000 years ago? _____

9. Who were the Hittites descended from? _____

10. How often were the Hittites mentioned in the King James Version of the Bible? _____

11. Who wrote the book *The Empire of the Hittites*? _____

12. In the Bible, how many references are there to Ur of the Chaldees? _____

13. Why did Woolley not excavate the cemetery as soon as he found it?_____

14. What was the name of the people who occupied ancient Ur? _____

15. Who discovered Nineveh? _____

16. What was the name of the ruins where Layard first started digging? _____

17. What was the name of the king of Israel that was mentioned on the black pillar Layard found in Nimrud?_____

18. What was the name of the king of Israel when Sennacherib besieged Jerusalem?_____

## Applied Learning Activity: (2 Points Each Blank)

19-21. Identify the writing materials and answer the questions: **Vellum, Papyrus, Pottery**

a. _____    b. _____    c. _____

22. What was vellum made from?_____
_____

23. What do you call a person who made vellum? _____

24. What was papyrus made from?_____
_____

25. Who made papyrus and sold it all over the Mediterranean? _____

26. What was the main city for papyrus production? _____

27. What word do we get from this city? _____

## Define: (5 Points Each Answer)

1. cuneiform: _____

2. strata: _____

3. syncline: _____

4. Persia: _____

5. rhyton: _____

6. cistern: _____

7. Nabataeans: _____

8. wadi: _____

## Short Answer Questions: (4 Points Each Question)

9. What was the name of the cuneiform record that told a story similar to the Bible record of Noah and the Flood? _____

10. Which Assyrian king compiled a library of tablets in Nineveh?_____

11. Which king made Babylon a city of gold?_____

12. Which Bible prophet predicted that Babylon would become uninhabited?_____

13. Who was the king who first carved out the Medo-Persian Empire?_____

14. In what year did he conquer Babylon? _____

15. Which Persian king left an inscription on the rock face of the Zagros Mountain near Bisitun?_____

16. What was the name of the great Persian city that Darius built?_____

17. Which Bible prophet wrote a book about Petra?_____

18. Which Roman emperor had a road made through Petra? _____

## Applied Learning Activity: (20 Points)

In your own words, tell the story of Esther. Include by name at least four of the characters and the name of the Jewish feast still celebrated today to commemorate the deliverance. (You may use the back of this page if more room is needed.)

## Define: (5 Points Each Answer)

1. Baal: _____

2. causeway: _____

3. Yehovah:_____

4. scroll: _____

5. annunciation: _____

6. Calvary: _____

7. Golgotha:_____

8. grotto: _____

9. Messiah: _____

10. ossuary: _____

## Multiple Answer Questions: (1 Point Each Blank)

11. What were the four main cities of ancient Phoenicia?

    a. _____     c. _____

    b. _____     d. _____

## Short Answer Questions: (4 Points Each Question)

12. What kind of trees did Solomon pay the Phoenician King Hiram for? _____

13. Whose tomb did Pierre Montet find? _____

14. Which Bible prophet challenged the prophets of Baal? _____

15. Which Bible prophet predicted that ancient Tyre would never be found? _____

16. In what year was the first Dead Sea Scroll found? _____

17. What was the name of the settlement near the cave where the Dead Sea Scrolls were found? _____
_____

18. Which Roman emperor adopted Christianity as the state religion?_____

19. Identify the languages: **Sumerian, Phoenician, Hebrew, Egyptian**

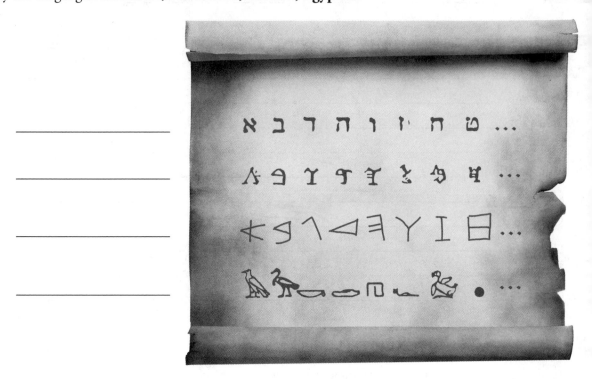

_____

_____

_____

_____

20. Name two books of the Bible that include chapters written in acrostic form (a form of Hebrew poetry):

a. _____

b. _____

| **T** | *The Archaeology Book*<br>Concepts & Comprehension | Test | Scope:<br>Chapters 1–11 | Total score:<br>____of 100 | Name |

## Define: (3 Points Each Answer)

1. carbon dating: _____

2. baulk: _____

3. synchronism: _____

4. mastabas: _____

5. centurion: _____

6. Chaldees: _____

7. bulla: _____

8. cuneiform: _____

9. syncline: _____

10. Persia: _____

11. rhyton: _____

12. annunciation: _____

13. ossuary: _____

14. grotto: _____

## Multiple Answer Questions: (1 Point Each Blank)

15. What are the four main periods of archaeological time?

   a. _____   c. _____

   b. _____   d. _____

16. What two nations did the Syrians think had come to attack them?

   a. _____   b. _____

17. What were the four main cities of ancient Phoenicia?

   a. _____   c. _____

   b. _____   d. _____

## Short Answer Questions: (3 Points Each Question)

18. Who was the first Egyptian to build a pyramid? _____

19. Who built the biggest pyramid in Egypt? _____

20. What was the name of the king of Israel that was mentioned on the black pillar Layard found in Nimrud? _____

21. What was the name of the king of Israel when Sennacherib besieged Jerusalem? _____

22. Which Bible prophet predicted that Babylon would become uninhabited? _____

23. Who was the king who first carved out the Medo-Persian Empire? _____

24. Which Bible prophet predicted that ancient Tyre would never be found? _____

25. In what year was the first Dead Sea Scroll found? _____

## Applied Learning Activity: (12 Points Total; 1 Point Each Answer)

26. Identify the Pyramids, Temples, Tombs, and unique features on Giza Map:

    Pyramid of Khufu

    Valley Temple of Khufu

    Pyramid of Menkaure

    Valley Temple of Menkaure

    Pyramid of Kahfre

    Valley Temple of Kahfre

    The Sphinx

    Temple of the Sphinx

    Pyramids of Queens

    Queen's Tombs

    Eastern Cemetery

    Mortuary Temple

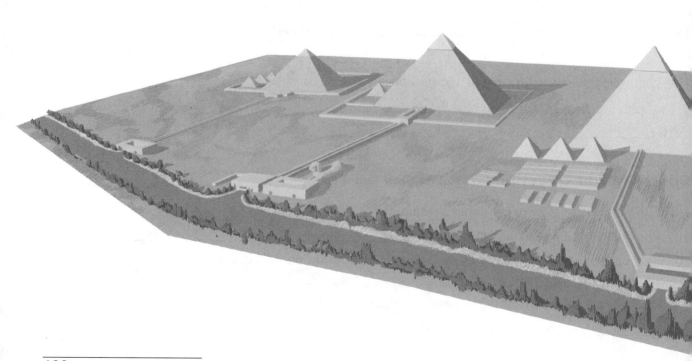

27. Identify the languages: **Sumerian, Phoenician, Hebrew, Egyptian**

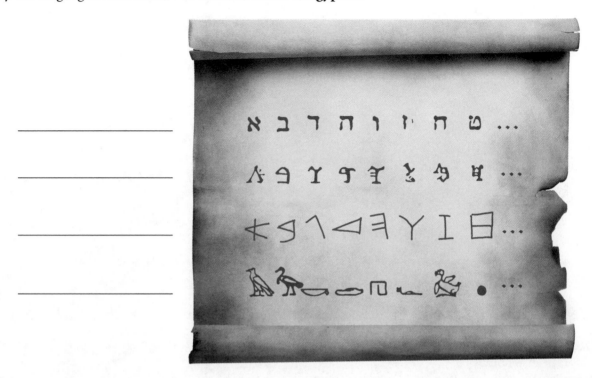

_____

_____

_____

_____

28. Name two books of the Bible that include chapters written in acrostic form (a form of Hebrew poetry):

a. _____

b. _____

## Define: (2 Points Each Answer)

. Principle of uniformity: _____

. Principle of catastrophe: _____

. Sediment: _____

. Metamorphism: _____

## Multiple Answer Questions: (3 Points Each Blank)

. There are two ways of thinking about the unobserved past. What are they? (3 Points Each Answer)

   a. _____

   b. _____

. In what order did God create the heavens and the earth?

   Day 1. _____    Day 4. _____

   Day 2. _____    Day 5. _____

   Day 3. _____    Day 6. _____

. What are the main "zones" into which the earth is divided?

   a. _____    c. _____

   b. _____    d. _____

. List the three types of plateaus and give an example of each.

   a. _____

   b. _____

   c. _____

. List the four types of mountains and give an example of each type.

   a. _____

   b. _____

   c. _____

   d. _____

## Applied Learning Activity: (5 Points Each Blank)

10. List which category the following types of rocks belong to.

Granite    a._____

Marble    b._____

Shale    c._____

Limestone  d._____

Coal    e._____

Rhyolite  f._____

Slate    g._____

## Define: (5 Points Each Answer)

1. Erosion: _____

2. Petrification: _____

3. Turbidite: _____

4. Gastrolith: _____

5. Fumaroles: _____

6. Carbon dating: _____

## Multiple Answer Questions: (2 Points Each Blank)

7. List the five primary causes of normal erosion.

   a. _____

   b. _____

   c. _____

   d. _____

   e. _____

8. Name four different types of fossils.

   a. _____     c. _____

   b. _____     d. _____

9. What are unstable atoms called? What is the most well-known radioactive atom?

   a. _____

   b. _____

## Short Answer Questions: (5 Points Each Question)

10. Sand is made primarily of what mineral? _____
    _____

11. How is a turbidite formed? _____
    _____

12. What is the main condition required for a fossil to form? _____
    _____

13. What factors influence whether a rock will bend or break? _____
    _____

**Applied Learning Activity:** (7 Points Each Question)

14. Describe how sedimentary rocks are formed.

15. Describe how metamorphic rocks are thought to form.

16. Explain the process of carbon dating in determining when a plant died.

17. In your own words, write a description of how dinosaur fossils were formed.

## Define: (5 Points Each Answer)

1. Magnetic field: _____

2. Uplift: _____

3. Second law of science: _____

4. Fountains of the deep: _____

5. Glacier: _____

6. Polar ice cap: _____

## Multiple Answer Questions: (2 Points Each Blank)

7. What four events have had the greatest impact in shaping the earth's geology?

    a. _____

    b. _____

    c. _____

    d. _____

8. What was the cause of the global Flood? What were some of the geological results of the Flood?

    a. _____

    b. _____

9. List the five ways to date the earth discussed in the book.

    a. _____

    b. _____

    c. _____

    d. _____

    e. _____

## Short Answer Questions: (5 Points Each Question)

10. What must we consider when evaluating conclusions obtained from these dating methods?_____
    _____

11. What role has the Fall played in shaping today's earth? _____
    _____

12. What conclusion can be drawn about the age of the earth from the various dating methods? _____
    _____

13. How did the creation event affect the earth's geology?_____
    _____

14. Why do we find ocean fossils near the top of Mt. Everest? _____

_____

15. What caused the Ice Age? _____

_____

## **Applied Learning Activity:** (18 Points)

16. Based on your previous essay, explain the difference between the secular view and the biblical view of the cause of the Ice Age.

## Define: (5 Points Each Answer)

1. Volcanism: _____

2. Escarpment: _____

## Short Answer Questions: (5 Points Each Question)

3. How was Grand Canyon formed?_____
_____

4. What causes the geysers in Yellowstone Park? _____
_____

5. How did Niagara Falls form? _____
_____

6. Why are the Appalachian and Rocky Mountains so different? _____
_____

7. How long does it take to form petrified wood?_____
_____

8. How are stalactites and stalagmites formed? _____

9. How is coal formed?_____
_____

10. How is natural gas formed? _____
_____

11. How is oil formed?_____
_____

12. Are dinosaur fossils the most abundant type of fossil?_____
_____

## Applied Learning Activity: (20 Points Each Question)

13. As part of God's judgment for disobedience at the end of time, the earth will undergo heat waves, droughts, flaming comets, earthquakes, and plagues. Read Psalm 46 and explain in your own words how this passage of Scripture explains the hope that Christians have in times of hardship on the earth.

14. Read 2 Peter 3:10–13 and Revelation 21:1–4. Based on these verses, explain in your own words the hope that Christians have for eternity. Where will you be throughout eternity?

| | The Geology Book — Concepts & Comprehension | Test | Scope: Chapters 1–8 | Total score: ____ of 100 | Name |

## Define: (2 Points Each Answer)

1. Principle of uniformity: _____

2. Principle of catastrophe: _____

3. Erosion: _____

4. Petrification: _____

5. Turbidite: _____

6. Gastrolith: _____

7. Fumaroles: _____

8. Metamorphism: _____

9. Magnetic field: _____

10. Sediment: _____

11. Second law of science: _____

12. Fountains of the deep: _____

13. Glacier: _____

14. Volcanism: _____

## Multiple Answer Questions: (2 Points Each Blank)

15. There are two ways of thinking about the unobserved past. What are they?

    a. _____

    b. _____

16. What are the main "zones" into which the earth is divided?

    a. _____

    b. _____

    c. _____

    d. _____

17. What four events have had the greatest impact in shaping the earth's geology?

    a. _____

    b. _____

    c. _____

    d. _____

## Short Answer Questions: (2 Points Each Question)

18. What is the main condition required for a fossil to form? _____

19. What factors influence whether a rock will bend or break? _____

20. What conclusion can be drawn about the age of the earth from the various dating methods? _____

21. How did the creation event affect the earth's geology? _____

22. Why do we find ocean fossils near the top of Mt. Everest? _____

23. What caused the Ice Age? _____

24. What was the cause of the global Flood? _____

25. How was Grand Canyon formed? _____

26. How did Niagara Falls form? _____

27. Why are the Appalachian and Rocky Mountains so different? _____

28. How long does it take to form petrified wood? _____

## Applied Learning Activity: (3 Points Each Answer)

29. List which category the following types of rocks belong to.

Granite    a._____

Marble    b._____

Shale    c._____

Limestone  d._____

Coal    e._____

Rhyolite   f._____

Slate    g._____

30. Describe how sedimentary rocks are formed.

31. Describe how metamorphic rocks are thought to form.

32. Explain the process of carbon dating in determining when a plant died.

## Define: (5 Points Each Answer)

1. karst: _____

2. Acheulean industry: _____

3. bas-reliefs: _____

4. Kyr: _____

5. Myr: _____

6. Neanderthals: _____

7. speleothems: _____

8. karst aquifer: _____

## Multiple Answer Questions: (2 Points Each Blank)

9. Name two large animals that inhabited caves prior to their extinction.

    a. _____

    b. _____

10. What are the three kinds of cave art that have been found?

    a. _____

    b. _____

    c. _____

## Short Answer Questions: (4 Points Each Question)

11. What is the probable reason some of our ancestors may have entered the "underland" of cave systems? ___
_____

12. What role did caves play for our ancestors?_____
_____

13. How much of the world's drinking water comes from limestone (karst) terrains? _____

14. How much is it estimated to be by 2025?_____

15. The strange event near the _____ not only split the once unified population, but also scattered those with different skills and abilities.

16. How did the events surrounding the Tower of Babel affect the ancient groups of people who disbursed from that area? _____
_____

17. When does the Bible mention caves for the first time? _____

18. In what country is Longgupo Cave, which is believed to host the oldest stone artifacts? _____
_____

19. What is the Twin River Cave in Zambia known for: (a) oldest human remains, (b) oldest burial site, or (c) art associated with burial rituals?

20. Were Neanderthals a different species than us? _____

## Applied Learning Activity: (10 Points)

21. Explain the relationship between cave paintings and acoustics and the conclusion the author draws from this relationship.

| **Q** | *The Cave Book* | Quiz 2 | Scope: | Total score: | Name |
|---|---|---|---|---|---|
| | Concepts & Comprehension | | Chapters 2–3 | ____ of 100 | |

## Define: (5 Points Each Answer)

1. troglobites: _____

2. troglophiles: _____

3. trogloxenes: _____

4. endogenetic: _____

5. exogenetic: _____

6. resurgences: _____

## Multiple Answer Questions: (2 Points Each Blank)

7. What is the longest period of cave habitation in modern times? Why did those humans choose to live inside a cave?

 a. _____

 b. _____

8. Name four civilizations that have caves present in their mythology.

 a. _____

 b. _____

 c. _____

 d. _____

9. Besides limestone, what other sedimentary rocks host many caves?

 a. _____

 b. _____

 c. _____

## Short Answer Questions: (4 Points Each Question)

10. Which is the largest troglobite alive today? _____

11. Are bats: (a) trogloxenes, (b) troglophiles, or (c) troglobites?

12. What is the Movile Cave in Romania famous for? _____

13. What are karstic rocks? _____

14. What percentage of the dry, ice-free landmass is covered by karstic rocks? _____

15. Where does the name "karst" come from? _____

16. What is a rhythmic spring?_____

17. What is the normal humidity inside most caves? _____
_____

## Applied Learning Activity: (2 Points Each Blank)

18. Movile Cave (Romania) (Fill in the blanks.)

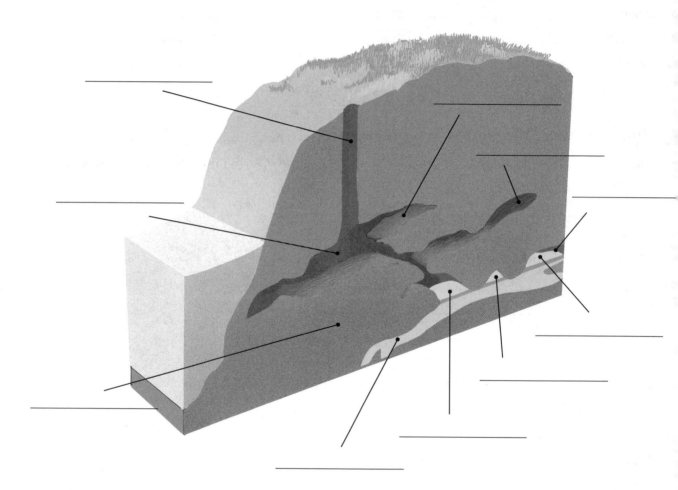

Manmade entrance shaft          Sulphurous water

Main passage                    Lake

Upper dry level                 Aired domepit 1 (23.3% oxygen)

59 ft.                          Aired domepit 2 (7.2% oxygen)

82 ft.                          Microbial mat

| Q | *The Cave Book*<br>Concepts & Comprehension | Quiz 3 | Scope:<br>Chapters 4–5 | Total score:<br>____of 100 | Name |

## Define: (6 Points Each Answer)

1. detrital formations:_____

2. dripping speleothems: _____

3. desiccation cracks: _____

4. master joints: _____

## Multiple Answer Questions: (2 Points Each Blank)

5. List the three types of active caves.

    a. _____

    b. _____

    c. _____

6. Name the two types of eccentric speleothems.

    a. _____

    b. _____

7. Name three non-calcite speleothems.

    a. _____

    b. _____

    c. _____

8. Name four pieces of gear needed for cave exploration.

    a. _____

    b. _____

    c. _____

    d. _____

9. Name four of the concerns, problems, and challenges faced by modern cave explorers as they study a cave.

    a. _____

    b. _____

    c. _____

    d. _____

10. What family spent the most time in a cave during World War II? a. _____

    What is the name of the cave system this family used? b. _____

11. What concerns are there for those trying to take photographs in a cave?

　　a. _____

　　b. _____

　　c. _____

## Short Answer Questions: (2 Points Each Question)

12. What is a shield?_____

13. What are cave rafts? _____

14. What is the most obvious and logical argument against a very old age of speleothems?_____
_____

15. Who was the first to build and use a diving device? _____

16. Who was the one who co-invented the Aqualung and was the first to SCUBA dive inside a cave? _____
_____

## Applied Learning Activity: (4 Points Each Blank)

17. Karst features diagram (Fill in the blanks.)

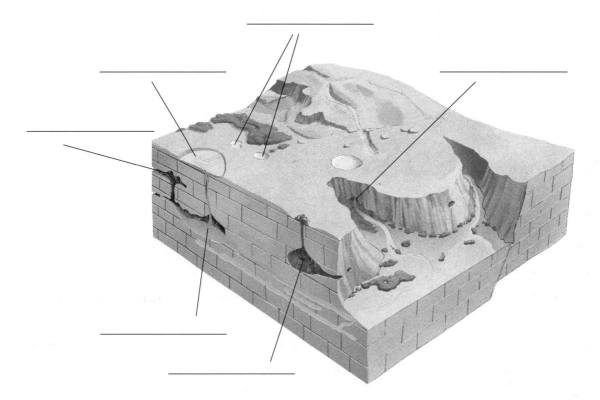

Sinkhole　　Fossil upper level　　Polje　　Active lower level　　Fossil cave　　Outflow cave

## Define: (5 Points Each Answer)

. cenote: _____

.. concavities: _____

. convexities: _____

. diagenesis: _____

. Xibalba: _____

## Multiple Answer Questions: (2 Points Each Blank)

. What are the two basic types of waters that reach karstic rocks?

    a. _____

    b. _____

. Extremely valuable information is yielded by continuous measurements of the discharge or flow rates (_____) and chemical composition (_____) of waters emerging through karst springs.

. Name five areas of science covered in the study of caves.

    a. _____

    b. _____

    c. _____

    d. _____

    e. _____

## Short Answer Questions: (4 Points Each Question)

. What is the only measurement of the karsting processes generally accepted?

0. How are karst aquifers viewed today?

11. Within what amount of time after the Flood did the Ice Age set in?

12. Why do karstologists never recommend any significant mining activities below the karst water table?

13. Discuss why karstlands are one of the most sensitive types of environments.

## Applied Learning Activity: (5 Points Each Answer)

14. Dead animals and man-made items left behind in caves are often covered with what? _____

15. How are these animals and items evidence for Creation? _____
_____

## Applied Learning Activity: (3 Points Each Blank)

16. List all of the sub-points for the three stages of cave formation based on the creationary model.

Stage 1:  a. _____

             b. _____

             c. _____

             d. _____

Stage 2:  e._____

             f. _____

             g. _____

             h._____

Stage 3:  i._____

## Define: (2 Points Each Answer)

1. bas-reliefs: _____

2. Kyr: _____

3. Myr: _____

4. troglobites: _____

5. troglophiles: _____

6. trogloxenes: _____

7. detrital formations:_____

8. dripping speleothems: _____

9. desiccation cracks: _____

10. concavities:_____

11. convexities:_____

12. diagenesis: _____

## Multiple Answer Questions: (2 Points Each Blank)

13. What are the three kinds of cave art that have been found?

    a. _____

    b. _____

    c. _____

14. Besides limestone, what other sedimentary rocks host many caves?

    a. _____

    b. _____

    c. _____

15. Name four pieces of gear needed for cave exploration.

    a. _____

    b. _____

    c. _____

    d. _____

16. Extremely valuable information is yielded by continuous measurements of the discharge or flow rates (_____) and chemical composition (_____) of waters emerging through karst springs.

17. The strange event near the _____ not only split the once unified population, but also scattered those with different skills and abilities.

18. Were Neanderthals a different species than us?

19. Are bats: (a) trogloxenes, (b) troglophiles, or (c) troglobites?

20. What is the Movile Cave in Romania famous for?

21. What is the most obvious and logical argument against a very old age of speleothems?

22. Who was the first to build and use a diving device?

23. Within what amount of time after the Flood did the Ice Age set in?

24. Why do karstologists never recommend any significant mining activities below the karst water table?

25. Movile Cave (Romania) (Fill in the blanks.)

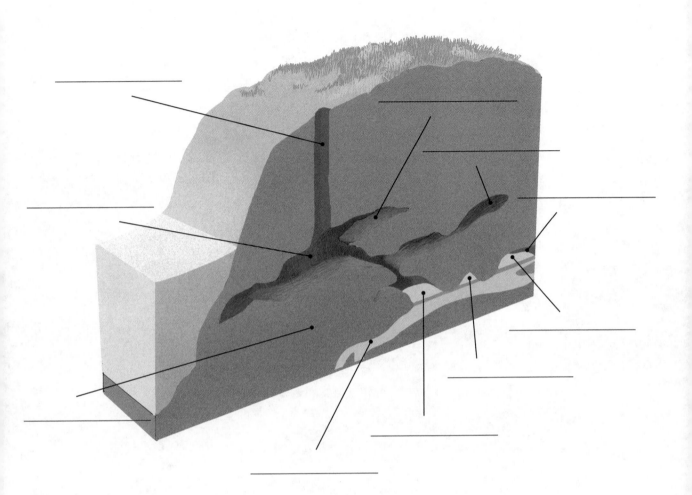

Manmade entrance shaft

Main passage

Upper dry level

59 ft.

82 ft.

Sulphurous water

Lake

Aired domepit 1 (23.3% oxygen)

Aired domepit 2 (7.2% oxygen)

Microbial mat

| **Q** | *The Fossil Book*<br>Concepts & Comprehension | Quiz 1 | Scope:<br>Intro–Ch 1 | Total score:<br>____of 100 | Name |

## Define: (5 Points Each Answer)

1. Evolution: _____

2. Paleontologist: _____

3. Permineralized fossils:_____

4. Trace fossils: _____

## Multiple Answer Questions: (2 Points Each Blank)

5. Who were the two men given credit for popularizing the modern teaching of evolution?

    a. _____

    b. _____

6. List two agents that erode and deposit sediments. Which agent is more powerful?

    a. _____

    b. _____

    c. _____

7. What two elements must exist in the right amount for sediments to turn into rocks?

    a. _____

    b. _____

8. What are the two most common rock cements?

    a. _____

    b. _____

9. Give three examples where calcium carbonate can be found.

    a. _____

    b. _____

    c. _____

## Short Answer Questions: (4 Points Each Question)

10. When did most of the branches of modern science begin?

11. What does TCSD stand for?

12. What types of sedimentary rocks are fossils normally found in?

13. What type of event would provide the right conditions to form fossils?

14. Briefly describe how a fossil can form.

15. Why must fossilization begin quickly?

16. What is the most common type of fossil?

17. What is the difference between permineralized wood and petrified wood?

18. Why is coal considered to be a fossil fuel?

19. Describe how coal forms (based on the research of Dr. Steve Austin).

20. What catastrophic event in May of 1980 supports Dr. Austin's theory?

21. How do polystrate fossils imply rapid burial?

## Applied Learning Activity: (1 Point Each Answer)

List and describe the four C's of biblical history discussed in this book.

22. a.             b.

23. a.             b.

24. a.             b.

25. a.             b.

## Define: (5 Points Each Answer)

1. Index fossils:_____

2. Geologic column:_____

3. Living fossils: _____

4. Trilobite: _____

5. Cambrian explosion: _____

6. Cavitations: _____

7. Paraconformities: _____

8. Stromatolites: _____

## Multiple Answer Questions: (2 Points Each Blank)

9. In what type of rock are most fossils found? Where can we see these layers of rock?

    a. _____

    b. _____

10. How many major geologic systems have been named? How many "super systems"?

    a. _____

    b. _____

11. Who said that fossils are "perhaps the most obvious and serious objection to the theory of evolution"? Why is this significant?

    a. _____

    b. _____

## Short Answer Questions: (4 Points Each Question)

12. According to Flood geologists, what does the geologic column show?

13. Explain why fossils of sea creatures are found throughout the geologic column while animals and land plants tend to be found higher in the column.

14. What tilted the basement rocks of the earth's surface?

15. How do the crystalline basement rocks found in Grand Canyon give testimony against millions of years of erosion?

16. Describe the "breached dam" concept for the formation of Grand Canyon.

17. What event supported the Flood geologists' interpretation of Grand Canyon's formation?

## Applied Learning Activity: (2 Points Each Answer)

18. Fossils are found in geologic systems (such as the Cambrian), somewhat as living things are found in ecological zones (such as the ponds and woodlands of the hardwood forest zone). Perhaps geologic systems or paleosystems are the remains of pre-Flood ecological zones. Identify the twelve Geologic Column systems represented using: Cambrian, Cretaceous, Devonian, Jurassic, Mississippian, Ordovician, Pennsylvanian, Permian, Quarternary, Silurian, Tertiary, Triassic

| **Q** | *The Fossil Book*<br>Concepts & Comprehension | Quiz 3 | Scope:<br>Chapter 4 | Total score:<br>____of 100 | Name |

## Define: (5 Points Each Answer)

1. Arthropod: _____

2. Cephalopods: _____

3. Diatoms: _____

4. Echinoderms: _____

5. Gastropods: _____

6. Malacology: _____

7. Nautiloids: _____

8. Palynology: _____

9. Protozoan: _____

10. Spicules: _____

## Multiple Answer Questions: (2 Points Each Answer)

11. What are the first animals fossilized in abundance? How does the complexity of these first fossils disprove evolution?

    a. _____

    b. _____

    _____

## Short Answer Questions: (4 Points Each Question)

12. Over 95% of all fossils found are _____.

13. How do fossil clams testify to a rapid burial?

14. What is perhaps the best mollusk evidence of creation and Flood geology?

15. What is the largest coral reef in the world?

16. Which layer of the geologic column is referred by evolutionists as the Age of Crinoids and Flood geologists as the Zone of Crinoids?

17. How do the crinoid fossils found in central United States testify to the Flood?

18. Insect fossils are rare since their hard outer layer decomposes easily after death. However, insect fossils with delicate details have been found. How is this possible?

19. How did fossils of sea creatures end up on the top of mountains?

## Applied Learning Activity: (2 Points Each Blank)

20-26: Triolobites (Fill in the blanks.)

20. _____

21. _____

22. _____

23. _____

24. _____

25. _____

26. _____

## Define: (6 Points Each Answer)

1. Evolutionary series: _____

2. Metamorphosis: _____

3. Splint bones: _____

4. Vertebrates: _____

## Multiple Answer Questions: (3 Points Each Blank)

5. What are the five groups of vertebrates?

    a. _____

    b. _____

    c. _____

    d. _____

    e. _____

6. Discuss two biblical explanations for the extinction of dinosaurs.

    a. _____
    _____

    b. _____
    _____

## Short Answer Questions: (4 Points Each Question)

7. Which of the following has the most DNA per cell of any other animal group?

    a. amphibians

    b. birds

    c. fish

    d. reptiles

    e. mammals

8. Coelacanths were once thought to be a transitional form between fish and amphibians. How was this claim disproved?

9. Explain what the Bible teaches about animals that eat meat today.

10. Discuss how dinosaurs can be explained by the biblical account of creation and the Flood.

11. What were the dimensions of the Ark?

12. What is likely about the dinosaurs on Noah's Ark during the Flood?

13. Why should *Archaeopteryx* not be considered a missing link?

14. Explain why the alleged sequence of horse hooves does not prove evolution.

## Applied Learning Activity: (6 Points Each Question)

15. Explain how the fossil record supports the biblical account of creation.

16. Explain how the fossil record supports the biblical account of the corruption of the earth.

17. Explain how the fossil record supports the biblical account of a global Flood.

## Short Answer: (5 Points)

18. Explain how the fossil record supports the biblical account of God's mercy on His creation.

## Define: (2 Points Each Answer)

1. Evolution: _____

2. Paleontologist: _____

3. Permineralized fossils:_____

4. Living fossils: _____

5. Trilobite: _____

6. Cambrian explosion: _____

7. Arthropod: _____

8. Cephalopods: _____

9. Diatoms: _____

10. Evolutionary series: _____

11. Metamorphosis: _____

12. Splint bones: _____

## Multiple Answer Questions: (2 Points Each Blank)

13. Who were the two men given credit for popularizing the modern teaching of evolution?

    a. _____

    b. _____

14. Who said that fossils are "perhaps the most obvious and serious objection to the theory of evolution"? Why is this significant?

    a. _____

    b. _____
    _____

15. What are the first animals fossilized in abundance? How does the complexity of these first fossils disprove evolution?

    a. _____

    b. _____
    _____

16. Discuss two biblical explanations for the extinction of dinosaurs.

    a. _____
    _____

    b. _____
    _____

17. What type of event would provide the right conditions to form fossils?

18. What catastrophic event in May of 1980 supports Dr. Austin's theory of how coal is formed?

19. According to Flood geologists, what does the geologic column show?

20. Explain why fossils of sea creatures are found throughout the geologic column while animals and land plants tend to be found higher in the column.

21. How do fossil clams testify to a rapid burial?

2. How did fossils of sea creatures end up on the top of mountains?

3. Why should *Archaeopteryx* not be considered a missing link?

4. Explain why the alleged sequence of horse hooves does not prove evolution.

List and describe the four C's of biblical history discussed in this book.

5. a.         b.

6. a.         b.

7. a.         b.

8. a.         b.

29. Fossils are found in geologic systems (such as the Cambrian), somewhat as living things are found in ecological zones (such as the ponds and woodlands of the hardwood forest zone). Perhaps geologic systems or paleosystems are the remains of pre-Flood ecological zones. Identify the twelve Geologic Column systems represented using: Cambrian, Cretaceous, Devonian, Jurassic, Mississippian, Ordovician, Pennsylvanian, Permian, Quarternary, Silurian, Tertiary, Triassic

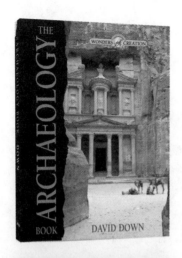
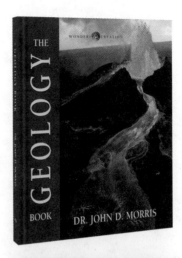
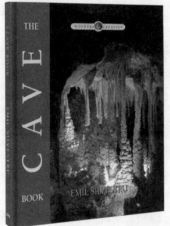
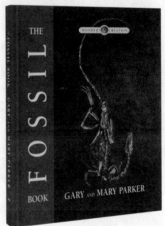

**Answer Keys**

**for Use with**

*General Science 2: Survey of Geology & Archaeology*

## Chapter 1 – What Archaeology Is All About – Worksheet 1

**accession year** — the year a king actually began his reign

**AD** — Anno Domini (the year of our Lord); the years after the Christian era began

**archaeiology** — study of beginnings

**artifact** — an item from antiquity found in an excavation

**BC** — Before Christ; the years before the Christian era began

**carbon dating** — calculating the amount of carbon left in organic material that has died

**ceramic** — something made of pottery

**chronology** — time periods; dates in which events happened

**debris** — discarded rubbish

**EB** — the Early Bronze Period

**exile** — a people sent out of their home country to another country

**exodus** — going out; applied to the Israelites leaving Egypt

**hieroglyphs** — Egyptian picture writing

**LB** — the Late Bronze Period

**MB** — the Middle Bronze Period

**millennium** — one thousand years

**non-accession year** — the first complete year of a king's reign

**pottery** — a vessel made of clay fired in a kiln

**synchronism** — something happening at the same time

**tell** — a Hebrew word meaning "ruins"; applied to hills on which people once lived

1. A study about beginnings
2. Defense, heat, and floods
3. 600 B.C.
4. It helps them identify from which period the pottery comes.
5. Early Bronze, Middle Bronze, Late Bronze, Iron Age

## Chapter 2 – Land of Egypt – Worksheet 1

**Asiatic** — in Egyptian terms, someone from Syria or Palestine

**baulk** — the vertical ridge left between two excavated squares in the ground

**dowry** — gift given to a prospective bride at the time of her marriage

**drachma** — a Greek coin worth about a day's wages

**dynasty** — a succession of kings descended from one another

**mastabas** — mud-brick structures beneath which were tomb chambers

**Nubia** — a country south of Egypt now called Sudan

**Pharoah** — title applied to many Egyptian kings

1. Misr
2. King Zoser's vizier, Imhotep
3. Khufu
4. Hapi
5. Straw

## Chapter 3 – The Hittites – Worksheet 1

**amphitheater** — a circle of seats surrounding an area where gladiators fought each other or fought wild beasts

**Anatolia** — mountainous area in central Turkey

**bathhouse** — a club where citizens could bathe in cold, warm, or hot water

**inscription** — writing made on clay, stone, papyrus, or animal skins

1. The Hittites
2. Hittites and Egyptians
3. Heth
4. Forty-six
5. William Wright

## Chapter 4 – Ur of the Chaldees – Worksheet 1

**centurion** — a military officer in charge of a hundred men

**Chaldees** — people who used to live in southern Iraq

**nomad** — a person who lived in a tent that could be moved from place to place

**papyrus** — sheets of writing material made from the Egyptian papyrus plant

1. Four
2. Sir Leonard Woolley
3. He wanted to learn more about Ur before he excavated such an important site.
4. Sumerians
5. Evidence of human sacrifice

## Chapter 5 – Assyria – Worksheet 1

**bulla** — an impression made on clay with a seal (plural: bullae)

**Medes** — people who used to live in northern Iran

**scarab** — model of a dung beetle with an inscription engraved on it for sealing documents

**seal** — an object made of stone, metal, or clay with a name engraved on it used to impress in soft clay

1. Henry Austin Layard
2. Nimrud
3. Jehu

4. Hezekiah

5. Forty-six

## Chapter 6 – Babylon: City of Gold – Worksheet 1

**Armenians** — people who lived in eastern Turkey and northern Iraq

**cuneiform** — a form of writing using a wedge-shaped stylus to make an impression on a clay tablet

**strata** — a layer of occupation exposed by excavations

**syncline** — a boat-shaped geological formation

1. The Gilgamesh Epic

2. Ashur-Bani-Pal

3. Asphalt

4. Nebuchadnezzar

5. Isaiah

## Chapter 7 – Persia – Worksheet 1

**Persia** — a country in central Iran

**rhyton** — a drinking vessel shaped like a human or animal

1. Cyrus the Great

2. 539 B.C.

3. Darius the Great

4. Persepolis

5. Haman

## Chapter 8 – Petra – Worksheet 1

**Bedouin** — Arabs living in tents with no fixed address

**cistern** — a hole dug in rock to store rainwater

**Edom** — country in southern Jordan

**Edomites** — people descended from Edom, also known as Esau, Jacob's brother

**Nabataeans** — people descended from Nabaioth, who occupied Petra

**siq** — narrow valley between two high rock formations

**theater** — a stage for actors in front of which was a semi-circle of seats

**wadi** — a dry riverbed, carrying water only when it rained

1. A.D. 1812

2. Esau's

3. Edomites

4. Obadiah

5. Trajan

## Chapter 9 – The Phoenicians – Worksheet 1

**Baal** — a word meaning "lord" and the name of a Phoenician god

**causeway** — a built-up road

**Yehovah** — a Hebrew name for God, usually spelled Jehovah, but there is no "J" in the Hebrew alphabet

1. Gebal, Berytus, Sidon, and Tyre
2. Cedars from Lebanon
3. Ahiram
4. Elijah
5. Ezekiel

## Chapter 10 – The Dead Sea Scrolls – Worksheet 1

**scroll** — papyrus or animal skin document rolled up into a cylinder

**vellum** — animal skin treated to be used as writing material

1. 1947
2. 22
3. Vellum
4. An acrostic
5. Qumran

## Chapter 11 – Israel – Worksheet 1

**annunciation** — an announcement

**Calvary** — Latin word meaning "skull"

**Golgotha** — Hebrew word meaning "skull"

**grotto** — cave

**Messiah** — meaning "Anointed One" and applied to an expected Jewish leader

**ossuary** — a box in which human bones were preserved

**Passover** — Jewish ceremony celebrating the Exodus from Egypt

1. Constantine
2. Yehovah saves
3. Jordan River
4. Capernaum
5. Skull

## Introduction & Chapter 1 – Planet Earth – Worksheet 1

**Principle of uniformity** — the scientific thought that past proccesses are no different than processes today, meaning everything happens by gradual processes over very long periods of time

**Principle of catastrophe** — the scientific thought that sees evidence of rapid, highly energetic events over short periods of time, doing a lot of geologic work

**Asthenosphere** — a suspected area in the uppermost portion of the earth's mantle where material is hot and deforms easily

**Plate** — huge regions of the earth identified by zones of earthquake activity

1. Origins science cannot be studied with repeatable, observable experiments in the present.

2. Uniformity (the present is the key to the past) and catastrophe (highly energetic events operated over short periods of time and did much geologic work rapidly)

3. In the Bible

4. Day 1: earth, space, time, light; Day 2: atmosphere; Day 3: dry land, plants; Day 4: sun, moon, stars, planets; Day 5: sea and flying creatures; Day 6: land animals, people

5. Sin can be defined as rebellion against God.

6. Crust, mantle, outer core, inner core

7. Continental crust (composed of granitic rock covered by sedimentary rock); oceanic crust (composed primarily of basaltic rock)

8. Provides the air we breathe, protects us from harmful cosmic radiation, and gives us weather

## Chapter 2 – The Ground We Stand Upon – Worksheet 1

**Igneous rock** — rock formed when hot, molten magma cools and solidifies

**Sedimentary rock** — rock formed by the deposition and consolidation of loose particles of sediment, and those formed by precipitation from water

**Metamorphic rock** — rocks formed when heat, pressure, and/or chemical action alters previously existing rock

**Ripple marks** — marks that indicate moving water flowed over a rock layer when the sediments were still muddy and yet to harden

**Crossbed** — areas of extremely large ripple marks

**Concretions** — concreted masses of sedimentary rock that has been eroded out of a softer area of rock

**Metamorphism** — a process of heat and pressure that causes one rock to alter into another

| Type | Category | Composition | Formation | Found |
|---|---|---|---|---|
| Granite | Igneous | Quartz and feldspar with mica and hornblende | Formed when molten rock is cooled | Mountains Upper mantle |
| Rhyolite | Igneous | Quartz and feldspar with mica and hornblende | Formed when molten rock erupts on land and solidifies | Land |

| Type | Category | Composition | Formation | Found |
|---|---|---|---|---|
| Obsidian | Igneous | Quartz and feldspar with mica and hornblende | Formed by the rapid cooling of lava as it flows on the surface of the ground | Land |
| Pumice | Igneous | Quartz and feldspar with mica and hornblende | Formed by eruptions on land—the cooling process forms air pockets in the rock | Land |
| Basalt | Igneous | Pyroxene, plagioclase feldspar | Solidified molten lava under water and on land | Oceanic crust, land |
| Shale | Clastic Sedimentary | Cemented particles of clay (and minor silt) | Formed from previously existing rocks, which were eroded, transported, and redeposited elsewhere | Mountains, land |
| Sandstone | Clastic Sedimentary | Quartz sand, particles big enough to be seen | Formed from previously existing rocks, which were eroded, transported, and redeposited elsewhere | Mountains, land |
| Conglomerate | Clastic Sedimentary | Pebble-size to boulder-size grains mixed with smaller sand or clay particles | Formed from previously existing rocks, which were eroded, transported, and redeposited elsewhere | Mountains, land |
| Limestone | Organic chemical sedimentary | Calcium carbonate from shells of sea creatures, reef fragments or limey secretions of sea creatures | Formed when water can no longer keep various chemicals dissolved within it | Sea floors, land |
| Diatomaceous earth | Organic chemical sedimentary | Collection of shells from diatoms or radiolarians and certain algae | Formed when water can no longer keep various chemicals dissolved within it | Land |
| Coal | Organic chemical sedimentary | Buried plant material | Formed when water can no longer keep various chemicals dissolved within it | Land |

| Type | Category | Composition | Formation | Found |
|---|---|---|---|---|
| Limestone | Inorganic chemical sedimentary | Calcium carbonate derived from inorganic sources | Formed when water can no longer keep various chemicals dissolved within it | Caves, mineral springs, stalactites, stalagmites |
| Dolomite | Inorganic chemical sedimentary | Calcium carbonate with magnesium atoms | Formed when water can no longer keep various chemicals dissolved within it | Land |
| Evaporites | Inorganic chemical sedimentary | The remains of evaporated seawater | Some were formed when a huge volume of mineral-laden water came up through the ocean floor basalts and released its dissolved content when it hit the cold ocean waters | Land |
| Slate | Metamorphic | Shale | Shale subjected to heat and pressure | Land |
| Schist | Metamorphic | Shale | Slate that continues to undergo heat and pressure | Land |
| Gneiss | Metamorphic | Alternating bands of different minerals from other sedimentary or igneous rocks | Formed from other sedimentary or igneous rocks that have been subjected to heat and pressure | Land |
| Quartzite | Metamorphic | Quartz sandstone | Quartz sandstone that has been subjected to change | Land |
| Marble | Metamorphic | Limestone | Heat and pressure applied to limestone | Land |

## Chapter 3 – The Earth's Surface – Worksheet 1

**Plain** — a broad area of relatively flat land

**Sediment** — a natural material broken down by processes of erosion and weathering; can be transported or deposited by water or wind

**Plateau** — flat lying sediment layers similar to plains but at higher elevations

**Mountain** — a large landform rising abruptly from the surrounding area

**Canyon** — a deep ravine between cliffs often carved by streams or rivers

1. Because the sediment deposited there is rich in nutrients

2. Fault: rock is broken and shoved up (Colorado Plateau); warped: regional squeezing or slow uplift (Appalachian Mountains); lava: hardened lava plains that may have been uplifted or hardened at the current level (Columbia River basalts)

3. Folded: layers of sediments that have been crumpled by pressures from the side (Alps, Himalayas, Appalachians, Rocky Mountains); domed: sediments pushed up from below (Black Hills of South Dakota); fault block: one area of sediments are pushed up (Grand Teton Mountains); volcanic: molten lavas pushed out to the surface of the earth (Hawaii's volcanic islands, Mount Rainier, Mount St. Helens, Mount Ararat)

4. Erosion

## Chapter 4 – Geological Processes and Rates – Worksheet 1

**Erosion** — the process by which soil and rock are worn away

**Deposition** — the process by which sediments are deposited onto a landform

**Turbidite** — an underwater rapid deposition of mud that hardens into a layer of rock formed by mud

1. Rain, ice, plants and animals, chemicals, ocean waves

2. Cavitation occurs when tiny bubbles in moving water explode inwardly; plucking is where rocks are picked up by moving water; kolk is like an underwater tornado that breaks up rock.

3. First, an event such as an earthquake starts a mud flow underwater. Next, the mud flow spreads out. Eventually the mud flow hardens into a layer of rock.

4. Flooding and tidal waves or tsunamis

## Chapter 4 – Geological Processes and Rates – Worksheet 2

**Compaction** — a process in the formation of sedimentary rock when the materials are pushed together tightly, leaving little to no open spaces

**Cementation** — a process in the formation of sedimentary rock when minerals are dissolved, which then help to solidify the rock by acting as glue or cement

**Fossils** — the remains of plants and animals that were once alive

**Petrification** — the process by which trees, plants, and even animals are solidified by burial in hot, silica-rich water

**Gastrolith** — rounded stones used by plant-eating dinosaurs to aid in digestion and sometimes found with fossilized remains

**Coprolite** — fossilized animal or dinosaur dung; can be used to determine a creature's diet

1. Silica

2. First, layers of sediment are deposited. The weight of the water and the sediments on top begin to compact the sediments underneath. Next, warm water circulates throughout the sediments and dissolves certain minerals. The dissolved minerals surround the individual grains of sediment. Finally, when the water cools off and stops moving, the dissolved minerals act as a "glue" that cements the grains of sediment together to form sedimentary rock.

3. The organism must be buried rapidly, protected from scavengers and from decomposition by bacteria and chemicals.

4. Hard parts are preserved; replacement by other minerals; cast or mold are all that remains; petrification; cabonization; preservation of soft parts; frozen animals; animal tracks and worm burrows; coprolites; gastroliths

5. First, if a dinosaur was not on the Ark, then it drowned in the Great Flood. Next, the animal was buried rapidly as the Flood deposited soft layers of material that later hardened into stone. Then, a process of fossilization occurred, such as the bones being replaced by dissolved minerals in the ground water. Finally, the fossils became exposed as the ground around the animal eroded away.

## Chapter 4 – Geological Processes and Rates – Worksheet 3

**Volcanism** — the eruption of molten rock (magma) onto the surface of the earth

**Fumaroles** — an opening in the earth's crust, usually associated with volcanic activity

**Geyser** — underground water that has been heated to an excessive degree and, because of pressure, bursts out of the ground temporarily

**Fault** — a fracture in rock along which separation or movement has taken place

**Earthquake** — a sudden release of energy below the earth's crust, which causes the earth's crust to move or shake

1. Some volcanoes erupt by just spilling lava out from their top; others explode out of their top.

2. In a normal fault, the hanging wall moves downward relative to the foot wall. In a reverse fault, the hanging wall moves upward. In a strike-slip fault, both walls move sideways.

3. Whether it is soft or brittle, how deep it is buried

4. Answers will vary.

## Chapter 4 – Geological Processes and Rates – Worksheet 4

**Atom** — the basic component of chemical elements

**Isotope** — variations of an element's atoms, usually in the different number of neutrons

**Radioisotope dating** — the process of using the rate of atomic decay to determine how old an object may be

**Carbon dating** — a process that uses the decay of carbon 14 to estimate the age of things that were once living

1. Heat and pressure recrystallize the minerals in rock into new mineral combinations. Some believe it happened over long periods of time; others believe it happened over short periods of time.

2. Radioactive; uranium

3. Uranium is used by nuclear power plants to generate electricity.

4. The daughter isotope is formed from the decay of the parent isotope.

5. When a plant is living, it takes the isotope carbon 14 into its leaves, stems, and seeds. After the plant dies, the carbon 14 decays into nitrogen 14. Scientists can measure the amounts of both carbon 14 and carbon 12. Since they know the time it takes the isotope to decay, they can calculate when the plant died.

## Chapter 5 – Ways to Date the Entire Earth – Worksheet 1

**Magnetic field** — a field that exerts forces on objects made of magnetic materials; made up of many lines of force

**Uplift** — in geology, a tectonic uplift is a geological process most often caused by plate tectonics, which causes an increase in elevation

1. Measuring the chemicals in the ocean, measuring the rate of erosion of the continents, measuring the sediments in the ocean, dating the atmosphere, dating the magnetic field

2. We must consider the possibility of processes happening at different rates. We can measure the rate that certain processes currently happen, but a massive flood or other event could have had a major impact in a very short amount of time.

3. A majority of methods used to age-date the earth yield ages far less than the acclaimed billions of years.

## Chapter 6 – Great Geologic Events of the Past – Worksheet 1

**Second law of science** — also referred to as the second law of thermodynamics, which states that in every process or reaction in the universe, the components deteriorate

**Fountains of the deep** — a phrase mentioned in Genesis 7 as a reference to sources of water as part of the Great Flood of Noah

**Glacier** — a huge mass of ice that moves slowly over land

**Polar ice cap** — a high latitude region of a planet that is covered in ice

1. Creation, the Fall, Flood, Ice Age

2. Formed the cores of the continents; some erosion and deposition probably happened

3. The Bible says in Genesis 3 that the entire creation came under the curse of sin, including plants, animals, mankind, and the earth. As a result of the curse, everything is wearing down and deteriorating.

4. God sent the Flood as a judgment on the wickedness of mankind. A worldwide flood would have caused a vast change to the earth's surface. The continents were uplifting, mountain ranges and lakes were formed, and rock and fossil layers were laid down.

5. The top of Mount Everest was once underwater and was later pushed up after the Flood waters receded.

6. See page 67 of *The Geology Book*.

## Chapter 7 – Questions People Ask – Worksheet 1

**Volcanism** — this is the process by which molten rock or lava erupts through the surface of a planet

**Escarpment** — a steep slope or long cliff that occurs from erosion or faulting and separates two relatively level areas of differing elevations

1. Many geologists now recognize that the Colorado River counld not have carved the Grand Canyon. At the end of Noah's Flood it appears that a great volume of water was trapped, held in place by the Kaibab Upwarp. Ice Age rains filled the lake to overflowing, and as it burst through its mountain "dam," the huge volume of lake waters carved the canyon.

2. At Yellowstone Park, the soil and rock is thin, allowing very hot material to be near the surface. As rain and run-off water trickle down into the earth, they get heated. In some places the underground water is trapped, and when heated to an excessive degree, it bursts out in a geyser. The geyser stops once the pressure is relieved, but will erupt again as building pressures exceed the maximum.

3. The level of water in Lake Erie is somewhat higher than the elevation of nearby Lake Ontario. A river draining the waters of Lake Erie into Lake Ontario runs over the Niagara escarpment, resulting in a spectacular set of falls. Erosion takes place as the water roars over the falls, and the escarpment naturally recedes toward Lake Erie.

4. Both mountain chains are the result of layers of sediments deposited by Noah's Flood. The Appalachians buckled up in the early stages of the Flood and were subjected to massive erosion by the continuing Flood waters. The Rocky Mountains buckled up late in the Flood, extending up above the Flood waters as the waters drained off. Thus, the erosion to which they were subjected was much less intense.

5. Petrified wood can form, under laboratory conditions, in a very short period of time. The speed of petrifaction is related to the pressures that inject the hot silica-rich waters into the wood.

6. When water saturated with calcium carbonate enters an open space such as a cave, it cools off or evaporates, leaving the calcium carbonate behind. Stalactites (holding "tightly" to the ceiling) and stalagmites (which are usually larger and thus more "mighty" than stalactites) are formed as this calcium carbonate precipitates out of the water.

7. Coal can be formed in a laboratory by heating organic material away from oxygen but in the presence of volcanic clay. Under such conditions, coal can be formed in a matter of hours. One wonders if the abundant forest growing before the Flood would not have formed huge log mats floating on the Flood ocean. As these decayed and were buried by hot sediments in the presence of volcanic clay, they might have rapidly turned to coal.

8. Natural gas, mostly methane, is given off in the coalification process. The largest quantities of it, however, are found in deep rocks not associated with decomposition of organic material. Evidently, some natural gas is from both organic and inorganic sources.

9. Many theories have been promoted as to the specific origin of oil. The best seems to be that it is the remains of algae once floating in the ocean but buried in ocean sediments. Oil is not the remains of dinosaurs as has sometimes been claimed.

10. No — they are rare. Most fossils are of sea creatures, fish, and insects. Relatively few fossils are of land animals, specifically dinosaurs.

## Introduction – Worksheet 1

**karst** — the term used by scientists to describe a landscape of caverns, sinking streams, sinkholes, and a vast array of small-scale features all generated by the solution of the bedrock, formed predominantly by limestones

**karst aquifers** — the assembly of ground water accumulated inside a karstic rock, enough to supply wells and springs

1. Their immediate need to find shelter from the rapidly cooling climate

2. Tower of Babel

3. 25 percent

4. Over 50 percent

1. It was deep inside the caves that some found shelter, mystical ritual hunting grounds, and a burial place for their dead.

2. The once-global knowledge and craftsmanship was split between many groups that could no longer truly communicate. Very quickly, various groups found themselves with the monopoly over one or several crafts/technologies, while other crafts were more or less lost for them. They were soon isolated from the other groups and many lost much of their knowledge of God.

## Chapter 1 – Humans and Caves – Worksheet 1

**acoustics** — points of resonance (locations where if certain musical notes are emitted, they will bounce back, amplified, from the walls)

**Acheulean industry** — from the town of Saint-Acheul, whose most characteristic tool was the stone hand axe

**bas-reliefs** — artwork usually made of soft, pliable clay attached to walls or even to large blocks

**cave paintings** — either simple outlines of charcoal or mineral pigment, or true paintings with outlines, shading, and vivid pigment fills

**engravings** — usually made on soft limestone surfaces

**Kyr** — abbreviation for thousand years

**Myr** — abbreviation for million years

**speleothems** — mineral deposits that form inside caves; especially stalagmites and stalactites

**Neanderthals** — believed by some to be early humans that were short, stocky, and stooped, with sloping foreheads, heavy eyebrows, jutting faces, and bent knees

1. Though we do not know for sure because there is no mention in Scripture, it is possible that there were caves prior to the Flood. They would have been formed differently than caves that exist today.

2. It is first mentioned in Genesis 19:30 concerning Lot and his daughters.

3. The word "cave" appears some 40 times in the Bible.

4. Cave bears, cave lions, and cave hyenas

5. China

6. Art associated with burial rituals

7. Paintings, engravings, and bas-reliefs

3. No, they were descended from the family of Noah.

1. Discussion might include their role as shelters or religious sanctuaries.

2. These early people carried their deep beliefs from their ancestor Noah. They also took on new beliefs as they separated from each other. Some may have come to see caves as an entrance into the earth. These were places of deep mystery to them.

3. The largest number of cave paintings are located in places of resonance (locations where if certain musical notes are emitted, they will bounce back, amplified, from the walls). It seems probable that chanting, dancing, and other types of ritual musical activities were associated with cave paintings.

4. The tools they made.

5. First representative of this human type was discovered in 1856 in a cave in the Neander Valley in Germany. Some have seen the remains as those belonging to an idiot, a hermit, or a medieval Mongolian warrior. Evolutionists were looking for a missing link, seeing this as a possible connection. However, they were simply humans with stocky, shorter bodies than many people today. They had broad noses and their brain size was slightly larger than that of modern humans.

6. Neanderthals had a spoken language, seemed to care for each other (those injured), and used flowers to decorate those buried.

## Chapter 2 – Caves and Mythology – Worksheet 1

**arthropods** — invertebrate animals having an exoskeleton, segmented body, and jointed appendages

**bidirectional air circulation** — air flowing two ways

**cul-de-sac** — cave with only one entrance

**echolocation** — bats send out sound waves that hit an object and an echo comes back, helping them identify the object

**troglobites** — creatures that live only in caves (from Greek for "cave dwellers")

**troglophiles** — creatures that spend some part of their life in caves (from Greek for "who like caves")

**trogloxenes** — creatures that got into a cave by accident and try to leave (from Greek for "foreign to caves")

**unidirectional air circulation** — air flowing one way

1. Egypt, Phoenicia, Assyro-Babylonia, Greece, Rome, and Maya

2. The cave olm

3. Troglophiles

4. A spectacular cave environment where several new species of creatures were found

5. Usually about 90 percent

1. Often one or two other females spread their wings underneath the delivering mother, ready to catch the little one if needed.

2. Thirty-eight Ukrainian Jews hid during World War II for nearly two years.

3. Ice can accumulate in cul-de-sac shafts because they act as traps for cold air.

4. Some caves have an abundance of negative ions in the air, which are usually oxygen atoms. Someone with a cold or flu can improve more quickly because of the absence of cosmic radiation.

# Chapter 3 – Caves and Karst – Worksheet 1

**cave** — considered a natural opening in rocks, accessible to humans, which is longer than it is deep and is at least 33 feet in length

**emergences** — the place where subterranean waters emerge to the surface

**endogenetic** — internal processes that can create caves

**exogenetic** — external process that can create caves

**karsted** — rich in karst features, especially caves

**orthokarst** — karst formed on carbonate rocks mainly by solution

**parakarst** — karst-like features formed on non-carbonate rocks, mainly by solution

**pseudokarst** — karst-like features formed on any kind of rock by other ways than solution

**resurgences** — the place where a sinking stream re-emerges to the surface

**sinkholes** — funnel-shaped hollows, from a few feet to hundreds of feet in diameter

1. Soluble rocks on which most landforms are formed by solution (karren, sinkholes, blind valleys, swallets, uvalas, poljes, etc.)

2. 12 percent

3. Evaporite rocks (rock gypsum and rock salt) and chalk

4. The exit point(s) of cave waters of a known stream, also called karst springs

5. Karst springs are called emergences when there is no evidence of the origin of the waters that emerge.

6. Up to 4,060.7 cubic feet per second; 115 tons of water every second, enough to supply the needs of more than two New Yorks every day.

7. Rhythmic springs flow intermittently, due to the very special shape and spatial distribution of the caves and conduits involved, as well as the constancy of water supply to the caves; Fontaine de Fontestorbes in southern France

1. Endogenetic caves are formed within moving lava. Lava tubes form when and where there are long-term lava flows. Often, stalactites also form. Exogenetic caves are the result of either chemical processes or physical processes, as volcanic ash and other pyroclastics are deposited.

2. Named by Austrian geographers and geologists in the 19th century while studying limestone terrains; they Germanized the Slovenian name "Kras" used by the locals. Probably comes from an old pre-Indo-European root "kara" meaning "desert of stone."

3. It might be riddled with all sorts of runnels, grooves, and small hollows called karren. Funnel-shaped hollows called sinkholes fill the terrain. Also, large hollows (depressions) called polje fill the karst terrain.

4. Northeastern South America (Venezuela and British Guiana); quartzite sandstone.

5. Many scientists agree today that this area was the result of the limestone being dissolved by sulfuric acid rising (not seeping down as in the case of proper karsting).


# Chapter 4 – Classifying Caves – Worksheet 1

**active caves** — live caves that have a flowing stream in them

**compoundrelict caves** — fossil caves above the water table

**denudation rate** — the pace at which a given surface of bare rock is eroded; usually measured in millimeters per millennium (thousand years)

**detrital formations** — sediments brought into the caves by streams and residual material left by the limestone

**dripping speleothems** — stalactites, stalagmites, and columns that are growing

**phreatic caves** — flooded (water saturated) caves formed and/or located below the water table

**relict caves** — caves without a flowing stream, which may have ponds or dripping water

**vadose caves** — caves that formed and continue to exist mostly above the water table; the majority of their passages have air above the water

1. A live cave that has a flowing stream

2. Three: inflow, outflow, and through caves

3. A special type of speleothem made up of two semicircular plates growing parallel to each other. They can grow more than three feet in diameter.

4. Two kinds; those that grow from stalactites (helictites) and those that grow from stalagmites (heligmites)

5. When cave pool water is saturated and very calm, thin flakes of calcite start growing, floating on the water, and are called cave rafts.

6. Any three: gypsum flowers, balloons, crusts, chandeliers, angel hair (mirabilite), moonmilk

7. Speleothems are nearly pure calcite that was removed from the limestone and redeposited inside caves. When removed from the limestone, some other soluble minerals accompany calcite, and some contain the radioisotope uranium 234. This decays through a long series of intermediates into lead 206. This is calculated based on a certain rate of decay.

8. Recorded growths of speleothems refute the necessity of tens or hundreds of thousands of years.

## Chapter 5 – Exploring Caves – Worksheet 1

**desiccation cracks** — cracks occurring because of shrinking of sediment as it dries

**master joints** — a tectonic discontinuity (fault line) that a given cave passage follows

**scallops** — spoon-shaped hollows in a cave wall, floor, or ceiling dissolved by eddies in flowing water

1. Alexander the Great in 325 B.C.

2. Jaques-Yves Cousteau

3. The Stermers (a Jewish family from Ukraine)

4. Popowa Yama Cave

1. Good footwear (hiking boots or rubber boots), good rope, flashlights, hard hat, gloves, pack, wool socks, and knee pads

2. This would include ascending and descending by rope, crawling through muck and water, moving through large and small chambers, following master joints and scallops.

3. Humidity increases the absorption of light; unpleasant colors can result when using artificial light films; much time is required to take simple shots with artificial lights.

## Chapter 6 – Studying Caves – Worksheet 1

**cenote** — steep-walled natural well reaching the water table and continuing below it

**concavities** — corresponding niches to convexities

**convexities** — a vertical succession of ledges

**diagenesis** — a complex set of transformations through which sediments go, from compaction, through dewatering to cementation

**Xibalba** — the Mayan "Underworld"

1.  The karst denudation rate
2.  The precipitation water and the fluvial water
3.  Hydrograph; chemograph
4.  As a transport and storage device
5.  About 500 years

1.  Subterranean geomorphology deals with the complex morphologies encountered under the ground and their relationship with the surface; geology studies the survey of all formations encountered to tectonics; geochemistry studies the direct chemical interactions of rocks with the environment; hydrology helps in understanding how waters move or are stored in the rocks; hydrogeology combines hydrology, geology, and chemistry.
2.  Water has a way of finding its way through karst aquifers, draining away from the artificial water reservoir.
3.  What happens on the surface has a significant effect on what happens under the ground because infiltration from surface water can be extremely fast.

## Introduction – Solving the Fossil Mystery – Worksheet 1

**Evolution** — the belief that life started by chance, and millions of years of struggle and death slowly changed a few simple living things into many varied and complex forms through stages

**Paleontologist** — a person who studies fossils

1. During the 1600s and 1700s

2. True

3. Charles Lyell and Charles Darwin

4. Time, chance, struggle, and death

5. Creation (God created all things in six actual days about 6,000 years ago. The completed creation was "very good"), corruption (Adam's sin ushered death, disease, sickness, pain, etc., into the world), catastrophe (God judged the wickedness of mankind with a global, earth-covering flood during Noah's day, around 4,500 years ago), and Christ (Jesus Christ came to earth to redeem mankind from the curse of sin and death).

## Chapter 1 – Fossils, Flooding, and Sedimentary Rock – Worksheet 1

**Archaeology** — the science that deals with human artifacts, and with things deliberately buried by humans

**Artifacts** — products crafted by humans

**Fossil** — remains or trace of a once-living thing preserved by natural processes, most often by rapid, deep burial in waterlaid sediments

**Geology** — the scientific study of the earth, including the materials that it is made of, the physical and chemical processes that occur on its surface and in its interior, and the history of the planet and its life forms

**Paleontology** — the study of fossils

**Permineralized fossils** — fossils preserved by minerals hardening in the pore spaces of a specimen such as a shell, bone, or wood

**Polystrates** — fossils that cut through many layers, suggesting the sequence was laid down very rapidly

**Pseudofossils** — false fossils; things that look like fossils but really aren't

**Sediments** — particles of sand, silt, clay, ash, etc., eroded and deposited by wind and water currents

**Trace fossils** — are not remains of plant or animal parts, but show evidence of once-living things

1. Flaky shale, gritty sandstone, or chalky limestone

2. Wind and water

3. Water

4. Water and rock cement

5. False. Rocks and fossils can form quickly given the right conditions. Long periods of time are not needed to form rocks and fossils.

6. Calcium carbonate and silica

7. Any three: limestone, bottom of tea kettle, in Tums and Rolaids, chalk

8. Silica gel packs are placed in boxes of electronic equipment.

9. A flood

10. When a plant is buried in sediment under flood conditions, the plant is preserved when the heavy sediment weight squeezes out extra water and encourages the growth of cement minerals that turn the plant into a fossil.

11. The plant or animal needs to be preserved quickly before it begins to decompose.

12. A permineralized fossil

13. Permineralized wood has minerals in its pore spaces but still has wood fibers, while minerals have completely replaced the wood but preserved the pattern in petrified wood.

14. Coal is the charred remains and carbon atoms of once-living plants, making it a fossil. Coal burns, making it a fuel.

15. Huge mats of vegetation were ripped up in violent storms, torn apart by the waves and currents, and deposited in layers. Sediment on top of these layers then squeezed out water and raised the temperature of the buried plants. The plants would then begin to char, turning into coal.

16. The eruption of Mount St. Helens

17. If the layers surrounding the polystrate item had built up slowly over millions of years, the tops of the polystrate item would rot away, even if the bottoms were fossilized.

## Chapter 2 – Geologic Column Diagram – Worksheet 1

**Geologic column** — a columnar diagram identifying rocky layers (strata) that form a sequence from bottom to top to indicate their relation to the twelve geologic systems

**Index fossil** — fossils used to identify a geologic system because they lived either (a) at a certain time or (b) in a certain place in the pre-Flood world

**Living fossils** — creatures found alive today that evolutionists thought became extinct millions of years ago

**Trilobite** — a crab-like creature that was the first fossil found buried in abundance around the world

1. Sedimentary rocks (limestone, shale, sandstone); cliffs, cuts, creeks, and quarries

2. Stages of evolutionary development over millions of years

3. A series of burials

4. 12; 3

5. Since they were buried later in Noah's Flood, paleosystems with land plants and animals occur higher in the geologic column diagram than those with only sea creatures, but fossils of sea life occur in all geologic systems or eco-sedimentary zones since the Flood waters eventually covered all the land.

6. Flood geologists use the word "first" to refer to the first to be buried by the Flood. They use the term "last" to refer to the last to be buried in the Flood. Evolutionists use the word "first" to refer to the first to evolve, meaning that nothing lived before it did. They use the word "last" to refer to the last surviving of its kind before it evolved into something else or became extinct.

7. Charles Darwin; Charles Darwin realized that evolution needed viable evidence of transitions from one animal into another; without them, evolution could not be validated.

## Chapter 3 – Flood Geology vs. Evolution – Worksheet 1

**Cambrian explosion** — the sudden appearance of a wide variety of complex life forms in the lowest rock layer with abundant fossils (Cambrian); considered a challenge to evolution, these may be the first organisms in a corrupted creation to be buried in Noah's Flood

**Cavitations** — bubbles formed by surging waters

**Paraconformities** — a gap without erosion in the geologic column diagram; breaks the time sequence assumed by evolution, and may suggest fossils from different environments were rapidly buried by a lot of water, not a lot of time

**Stromatolites** — banded rock deposits formed by blue-greens growing in mossy mats on rocks in the tidal zone along the shore; the mats trap and then cement sand grains to form a mineral layer, continually building new layers on top of earlier ones

1. Cambrian

2. They tried to look for simple life forms in pre-Cambrian rock.

3. False. Jellyfish and segmented worms are anything but simple.

4. Two-thirds

5. The waters that burst out of the deep during Noah's Flood

6. If the millions-of-years scenario were true and erosion occurred gradually, the softer rock would be gone and the hard rock would stick up into the sediment above. However, the tilted layers have been sheared off in a nearly straight line.

7. The bottom layers were most likely formed in the years before the Flood and sheared off during the beginning stages of the Flood. The upper layers were set into place during the Flood. Tectonic activity pushed these layers up as the water receded from the earth's surface during the later stages of the Flood. Water became trapped by earthen dams, which finally broke years after the Flood and released water to tear away the earth's surface. These cascades of water followed the easiest path downhill, which is now where we see the Colorado River through the gorge of Grand Canyon.

8. Mount St. Helens

## Chapter 4 – Kinds of Fossils I: Invertebrates – Worksheet 1

**Arthropod** — all creatures with jointed legs and a tough outside skeleton (exoskeleton) made of chitin: insects, crabs and shrimp, spiders, centipedes, and millipedes

**Cephalopods** — means "head-footed," since their tentacles come out of their heads; the most complex of all the invertebrates are the squid and octopus in the mollusk class

**Diatom** — microscopic, one-celled plants whose walls are decorated with glass ($SiO_2$) in exquisite patterns; mined and sold as diatomaceous earth, which is used in filtering and abrasion

**Echinoderms** — meaning "spiny-skinned," members of the starfish/sea star group usually have bony plates and spines embedded

**Gastropods** — means "stomach-footed," since they walk on their stomachs; mollusk class to which snails belong

**Invertebrate** — animals without backbones

**Malacology** — a branch of science devoted to the study of mollusk shells

**Mollusks** — a large phylum of animals with thick, muscular bodies and a complex system of organs

**Nautiloids** — fossils with tapered, chambered shells; some are coiled like the modern nautilus, others are curved like bananas, and still others are straight, like ice cream cones

**Palynology** — the branch of paleontology that studies microscopic spores and pollen of plants

**Protozoan** — one-celled animal

**Spicules** — sponges that have hard skeletal structures of crystal-like spines

1. Seashells

2. True

3. Many fossil clams are found with both sides of the shell still together. That means the clam must have been buried so deeply and so fast that it couldn't even open its shell to burrow out.

4. Shelled squids

5. The sutures of nautiloids vary stratigraphically from smooth to wiggly to very wiggly. Evolutionists use this to "prove" simple to more complex lifeforms.

6. First, the series starts with a very complex animal at the bottom of the GCD. Second, the first-buried form is also the fittest, since it's the only survivor. Third, the animal never evolved from anything or into anything. Fourth, there are few suggestions and no agreement on the survival value of having a wiggly suture. Fifth, there are reversals of the sequence evolutionists expected.

7. Great Barrier Reef in Australia

8. False. Scientists have measured coral growth and found that the largest reef in the world could have formed in less than 4,000 years.

9. Mississippian layer

10. They show that those areas were at one time completely covered by water.

11. It is possible that the volcanic activity that accompanied the Flood released toxins into the water that prevented the decomposition of the insects. Silt and clay buried the insects and settled quickly in briefly quiet water, solidifying fast enough to prevent later currents from tearing apart the fragile specimens. Insects can also be preserved in detail in hardened pine sap (amber).

12. Trilobites

13. Evolution assumes that the earliest fossils, which are found in the lowest layers, would be the most primitive and least complex since they hadn't yet evolved into more complex beings. However, since these fossils reveal complex creatures of design, they disprove the idea that non-complex beings changed into complex beings.

14. The turbulent Flood waters covered the entire earth, including the high hills (Genesis 7:19). Then the mountains rose, and the valleys sank down (Psalm 104:8). At the end of the Flood, God raised up the layers that were below the sea, lifting sea-creature fossils even to the tops of earth's highest mountains.

**Chapter 5 – Kinds of Fossils II: Vertebrates – Worksheet 1**

**Evolutionary series** — a sequence of fossils that suggests how one kind of creature might have changed into another

**Metamorphosis** — the process of transformation from an immature form to an adult form in two or more distinct stages

**Splint bones** — modern one-toed horses actually keep parts of the two flanking toes as important leg support structures (not useless evolutionary leftovers)

**Vertebrates** — animals with backbones

1. Fish, amphibians, reptiles, birds, and mammals

2. a. amphibians

3. Live coelacanths were found in the Indian Ocean and near Indonesia. There were no elbow or wrist joints as evolutionists once claimed; the stiff fin was used for steering and swimming, not walking. Their organs worked more like those in a shark, not those in a frog. The fish did not live in ponds; it lived in the deep ocean.

4. God created all animals and people in the beginning to eat only plants. It was only after man's sin corrupted God's perfect creation that some animals began to eat other animals, and it was only after the

Flood that God gave mankind permission to eat meat.

5. Dinosaurs were created on Day 6 along with the other land animals and people. Two representatives of the various dinosaur kinds were on the Ark. Since the average height of dinosaurs was about the size of a small pony, and since younger dinosaurs were smaller than older ones, they would have fit on Noah's Ark during the global Flood. Those that weren't on the Ark perished in the Flood. Many were buried in the muddy sediments. Those that survived the Flood on the Ark repopulated the earth after disembarking, although most eventually died from various causes.

6. 450 feet long; 75 feet wide; 45 feet high (150 x 25 x 15 meters)

7. They were likely young adults since God desired them to replenish the earth after the Flood.

8. The first explanation is because climate and soil conditions changed; dinosaurs had a difficult time surviving in that "new" world. The second explanation is that they were over-hunted by people after the Flood. Both are certainly possibilities, but we are not absolutely certain why dinosaurs went extinct.

9. Features in *Archaeopteryx* that evolutionists claimed to be transitional were and are found in other birds. The conclusion was made that *Archaeopteryx* was just a strong flying bird. Additionally, regular birds are found lower in the geologic column than *Archaeopteryx*.

10. First, the claimed earliest horse fossil is actually still seen today as a hyrax, or coney, not a primitive horse. Second, the change in horse size is not support for evolution since extreme sizes can be achieved simply by breeding. Third, the difference in hoof number can be explained by variation in horse kinds that were better able to live in different environments. Finally, the three-hoofed and one-hoofed horse kinds lived at the same time. These animals are not evidence for evolutionary transition.

11. Answers will vary.

## Conclusion – Worksheet 1

1. The first buried fossils of each group are complete and complex, with all the features that separate its kind from all the others.

2. Since Adam sinned, the earth and all that was in it was cursed. Fossils themselves are dead things. Death was not part of God's original creation; it came as a result of sin. The fossil record also testifies to animals eating other animals; this did not occur before sin since there was no death before sin and since God originally created man and animals to be vegetarian.

3. 1. Dead things are broken down so fast that most fossils must have formed rapidly or they wouldn't have formed at all.

   2. Most fossils are found in sedimentary rocks that form in the way concrete cures, so the right conditions form rock quickly and no amount of time can form rocks under the wrong conditions.

   3. Some dinosaur bones and other fossils contain DNA, protein, or other chemicals that would break down completely in thousands of years, not millions.

   4. Countless numbers of living things must have been buried at the same time and place to form oil deposits, and that must have happened no more than thousands of years ago, or the oil would have leaked to the surface.

   5. Gaps in the GCD with insufficient evidence of erosion, such as the "150 million missing years" in the walls of Grand Canyon, suggest evolution's millions of years are a myth.

   6. Misplaced fossils, like fossils of woody plants in Cambrian rock and living fossils, show that fossils from various geologic systems lived at the same time in different places, not at different times in the same place. The systems in the geologic column seem to be primarily the buried remains of different life zones in the pre-Flood world.

7. Scientists studying the 1980 and 1982 eruptions of Mount St. Helens saw powerful evidence that catastrophic processes can do in days what slow processes could never do, not even in millions of years.

4. God preserved His creation while enacting judgment on the world. Noah, his wife, his sons, and their wives, along with two of every kind of unclean land animal and seven of every kind of clean land animal, were preserved by God's grace on the Ark. He promised judgment and yet made a way to escape that judgment. After the Flood, God restored His creation to life again. Man reproduced after his kind, animals reproduced after their kind, and plants reproduced after their kind, just as God commanded and desired.

## Application – Worksheet 1

1. Sedimentary rocks – limestones, sandstones, shale

2. Cliffs, cuts (for roads), creeks, quarries

3. No — some states and specific locations like national parks require special permits; others will allow you to look for fossils but you cannot collect them.

4. Always check with your state laws, but usually in the US, property owners are considered the owners of any fossils found on their property.

5. A lot of rock removal has already been done, and sometimes you can see fossils that have been exposed or easily find them in the rock remains from quarrying.

6. Quarries can be very unstable and contain quicksand, rockfalls, deep water, sharp rocks, dangerous cliffs, and even dangerous creatures like snakes or alligators.

7. Answer will vary; can include box, bag, bucket, tissue, cloth, sealable or plastic bags

8. Wet screening is when a screen is used with a shovel or scoop that allows you to pour the sandy material onto the screen and then gently shake it in the water; this works best at a beach or even without water in sandy locations on land

9. Matrix

10. Answers will vary. Fossils are old and fragile — you could easily damage the fossil with the wrong tools or by getting too close to it.

11. A. Trilobites,  B. Echinoderms,  C. nautiliods,  D. corals,  E. Brachiopods,  F. ferns,  G. mollusks

12. By matching the size scale and other identifying visual features, you can begin to visually identify fossils in the field. You also may not recognize a fossil by scientific name but you can by sight.

13. No. Most fossils need only minimal cleaning; fragile or crumbly fossils can be stabilized by using a concrete sealer.

14. Match the fossil with the biblical application by drawing lines:
    a. Nautilus > Creation
    b. Trilobites or fossils with bite marks > Corruption
    c. Living Fossils > Christ
    d. Closed fossil clams > Catastrophe

Bonus Question: Answers will vary.

True/False Questions:

1. F,  2. F,  3. F,  4. T,  5. F,  6. F,  7. T,  8. F,  9. T,  10. F,  11. T,  12. F,  13. T,  14. T,

15. F (many show complex beginning like trilobites and nautiloids)

## Unit One Quiz, chapters 1–2

1. **accession year** — the year a king actually began his reign
2. **AD** — Anno Domini (the year of our Lord); the years after the Christian era began
3. **BC** — Before Christ; the years before the Christian era began
4. **carbon dating** — calculating the amount of carbon left in organic material that has died
5. **EB** — the Early Bronze Period
6. **LB** — the Late Bronze Period
7. **MB** — the Middle Bronze Period
8. **baulk** — the vertical ridge left between two excavated squares in the ground
9. **synchronism** — something happening at the same time
10. **mastabas** — mud-brick structures beneath which were tomb chambers
11. Early Bronze, Middle Bronze, Late Bronze, Iron Age
12. Defense, heat, and floods
13. A study about beginnings
14. 600 B.C.
15. Hapi
16. Misr
17. King Zoser's vizier, Imhotep
18. Khufu
19. Identify the Pyramids, Temples, Tombs, and unique features on Giza Map:

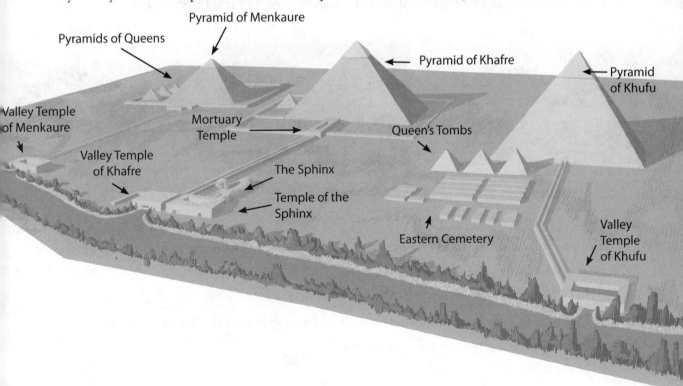

**Unit Two Quiz, chapters 3–5**

1. **amphitheater** — a circle of seats surrounding an area where gladiators fought each other or fought wild beasts

2. **Anatolia** — mountainous area in central Turkey

3. **centurion** — a military officer in charge of a hundred men

4. **Chaldees** — people who used to live in southern Iraq

5. **bulla** — an impression made on clay with a seal (plural: bullae)

6. **scarab** — model of a dung beetle with an inscription engraved on it for sealing documents

7. The Hittites and Egyptians

8. The strongest nation in the Middle East 3000 years ago was the Hittites.

9. The Hittites were descended from Heth.

10. The Hittites were mentioned forty-six times in the King James Version of the Bible.

11. William Wright wrote the book *The Empire of the Hittites.*

12. In the Bible, there are four references to "Ur of the Chaldees."

13. He wanted to learn more about Ur before he excavated such an important site.

14. Sumerians was the name of the people who occupied ancient Ur.

15. Henry Austin Layard discovered Nineveh.

16. Nimrud was the name of the ruins where Layard first started digging.

17. Jehu was the name of the king of Israel that was mentioned on the black pillar Layard found in Nimrud.

18. Hezekiah was the name of the king of Israel when Sennacherib besieged Jerusalem.

19-21. Identify the writing materials and answer the questions: **Vellum, Papyrus, Pottery**

a. Pottery          b. Vellum          c. Papyrus

22. Vellum was made from leather (animal skins that were scraped clean and treated for preservation).

23. A person who made vellum was called a tanner.

24. Papyrus was made from papyrus stalks from Egypt.

25. The Phoenicians, now Lebanese, made papyrus and sold it all over the Mediterranean.

26. Byblos was the main city for papyrus production.

27. We get the word "Bible" from this city.

**Unit Three Quiz, chapters 6–8**

1. **cuneiform** — a form of writing using a wedge-shaped stylus to make an impression on a clay tablet

2. **strata** — a layer of occupation exposed by excavations

3. **syncline** — a boat-shaped geological formation

4. **Persia** — a country in central Iran

5. **rhyton** — a drinking vessel shaped like a human or animal

6. **cistern** — a hole dug in rock to store rainwater

7. **Nabataeans** — people descended from Nabaioth who occupied Petra

8. **wadi** — a dry river bed, carrying water only when it rained

9. The Gilgamesh Epic

10. Ashur-Bani-Pal

11. Nebuchadnezzar

12. Isaiah

13. Cyrus the Great

14. 539 B.C.

15. Darius the Great

16. Persepolis

17. Obadiah

18. Trajan

19. **Applied Learning Activity:** (20 Points - 4 Points for each character and for Purim)

    Student should include by name at least four of the characters (Darius, Xerxes, Vashti, Esther, Haman, Mordecai) and describe their role in the account. The story should reflect the biblical account of Esther.

    Purim is the name of the Jewish feast still celebrated today to commemorate the deliverance.


**Unit Four Quiz, chapters 9–11**

1. **Baal** — a word meaning "lord" and the name of a Phoenician god

2. **causeway** — a built-up road

3. **Yehovah** — a Hebrew name for God, usually spelled Jehovah, but there is no "J" in the Hebrew alphabet

4. **scroll** — papyrus or animal skin document rolled up into a cylinder

5. **annunciation** — an announcement

6. **Calvary** — Latin word meaning "skull"

7. **Golgotha** — Hebrew word meaning "skull"

8. **grotto** — cave

9. **Messiah** — meaning "Anointed One" and applied to an expected Jewish leader

10. **ossuary** — a box in which human bones were preserved

11. Gebal, Berytus, Sidon, and Tyre

12. Cedars from Lebanon

13. Ahiram

14. Elijah

15. Ezekiel

16. 1947

17. Qumram

18. Constantine

19.

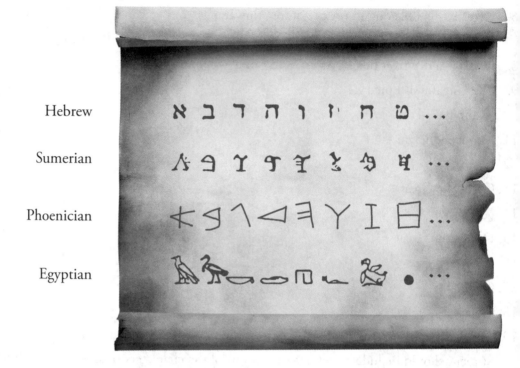

19. a. Psalms

   b. Lamentations

1. **carbon dating** — calculating the amount of carbon left in organic material that has died
2. **baulk** — the vertical ridge left between two excavated squares in the ground
3. **synchronism** — something happening at the same time
4. **mastabas** — mud-brick structures beneath which were tomb chambers
5. **centurion** — a military officer in charge of a hundred men
6. **Chaldees** — people who used to live in southern Iraq
7. **bulla** — an impression made on clay with a seal (plural: bullae)
8. **cuneiform** — a form of writing using a wedge-shaped stylus to make an impression on a clay tablet
9. **syncline** — a boat-shaped geological formation
10. **Persia** — a country in central Iran
11. **rhyton** — a drinking vessel shaped like a human or animal
12. **annunciation** — an announcement
13. **ossuary** — a box in which human bones were preserved
14. **grotto** — cave
15. Early Bronze, Middle Bronze, Late Bronze, Iron Age
16. The Hittites and Egyptians
17. Gebal, Berytus, Sidon, and Tyre
18. King Zoser's vizier, Imhotep
19. Khufu
20. Jehu
21. Hezekiah
22. Isaiah
23. Cyrus the Great
24. Ezekiel
25. 1947

26. Identify the Pyramids, Temples, Tombs, and unique features on Giza Map:

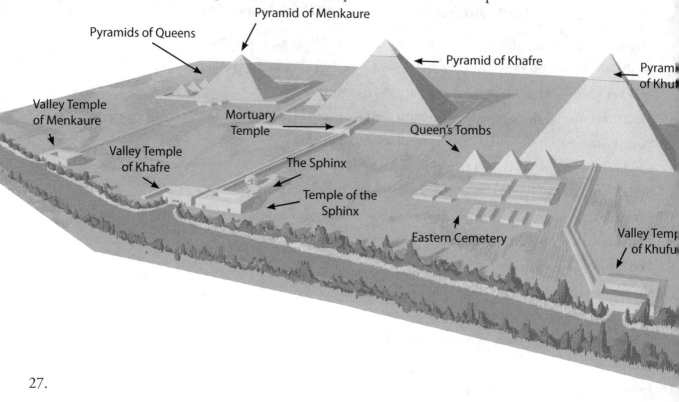

27.

Hebrew

Sumerian

Phoenician

Egyptian

28. a. Psalms

   b. Lamentations

## Unit One Quiz, chapters 1–3

1. **Principle of uniformity** — the scientific thought that past proccesses are no different than processes today, meaning everything happens by gradual processes over very long periods of time

2. **Principle of catastrophe** — the scientific thought that sees evidence of rapid, highly energetic events over short periods of time, doing a lot of geologic work

3. **Sediment** — natural materials broken down by processes of erosion and weathering; can be transported or deposited by water or wind

4. **Metamorphism** — a process of heat and pressure that causes one rock to alter into another

5. Uniformity (the present is the key to the past) and catastrophe (highly energetic events operated over short periods of time and did much geologic work rapidly)

6. Day 1: earth, space, time, light; Day 2: atmosphere; Day 3: dry land, plants; Day 4: sun, moon, stars, planets; Day 5: sea and flying creatures; Day 6: land animals, people

7. Crust, mantle, outer core, inner core

8. Fault (Colorado Plateau), Warped (Appalachians), Lava (Columbia River basalts)

9. Folded (Alps, Himalayas, Appalachians, Rocky Mountains), Domed (Black Hills of South Dakota), Fault block (Grand Teton Mountains), Volcanic (Hawaii's volcanic islands, Mount Rainier, Mount St. Helens, Mount Ararat)

10. a. Granite: Igneous, b. Marble: Metamorphic, c. Shale: Classic Sedimentary, d. Limestone: Organic Chemical Sedentary, e. Coal: Organic Chemical Sedentary, f. Rhyolite: Igneous, g. Slate: Metamorphic

## Unit Two Quiz, chapter 4

1. **Erosion** — states if processes continue as they do today, everything will eventually be eroded or worn away

2. **Petrification** — the process by which trees, plants, and even animals are solidified by burial in hot, silica-rich water

3. **Turbidite** — an underwater, rapid deposition of mud that hardens into a layer of rock

4. **Gastrolith** — rounded stones used by plant-eating dinosaurs to aid in digestion and sometimes found with fossilized remains

5. **Fumaroles** — an opening in the earth's crust, usually associated with volcanic activity

6. **Carbon dating** — a process that uses the decay of carbon 14 to estimate the age of things that were once living

7. Rain, ice, plants and animals, chemicals, ocean waves

8. Any four: hard parts are preserved; replacement by other minerals; cast or mold are all that remains; petrification; cabonization; preservation of soft parts; frozen animals; animal tracks and worm burrows; coprolites; gastroliths

9. Radioactive; uranium

10. Silica

11. First, an event such as an earthquake starts a mud flow underwater. Next, the mud flow spreads out. Eventually the mud flow hardens into a layer of rock.

12. The organism must be buried rapidly, protected from scavengers and from decomposition by bacteria and chemicals.

13. Whether it is soft or brittle, how deep it is buried

14. First, layers of sediment are deposited. The weight of the water and the sediments on top begin to compact the sediments underneath. Next, warm water circulates throughout the sediments and dissolves certain minerals. The dissolved minerals surround the individual grains of sediment. Finally, when the water cools off and stops moving, the dissolved minerals act as a "glue" that cements the grains of sediment together to form sedimentary rock.

15. Heat and pressure recrystallize the minerals in rock into new mineral combinations. Some believe it happened over long periods of time; others believe it happened over short periods of time.

16. When a plant is living, it takes the isotope carbon 14 into its leaves, stems, and seeds. After the plant dies, the carbon 14 decays into nitrogen 14. Scientists can measure the amounts of both carbon 14 and carbon 12. Since they know the time it takes the isotope to decay, they can calculate when the plant died.

17. Answers will vary but should include key ideas: First, if a dinosaur was not on the Ark, then it drowned in the Flood. Next, the animal was buried rapidly as the Flood deposited soft layers of material that later hardened into stone. Then, a process of fossilization occurred, such as the bones being replaced by dissolved minerals in the ground water. Finally, the fossils became exposed as the ground around the animal eroded away.

## Unit Three Quiz, chapters 5–6

1. **Magnetic field** — a field that exerts forces on objects made of magnetic materials; made up of many lines of force

2. **Uplift** — in geology, a tectonic uplift is a geological process most often caused by plate tectonics, which causes an increase in elevation

3. **Second law of science** — also referred to as the second law of thermodynamics, which states that in every process or reaction in the universe, the components deteriorate

4. **Fountains of the deep** — a phrase mentioned in Genesis 7 as a reference to sources of water as part of the Great Flood of Noah

5. **Glacier** — a huge mass of ice that moves slowly over land

6. **Polar ice cap** — a high latitude region of a planet that is covered in ice

7. Creation, the Fall, Flood, Ice Age

8. God sent the Flood as a judgment on the wickedness of mankind. The Flood formed many of the rock and fossil layers. See pages 63–66 for more details.

9. a. chemicals in the ocean, b. erosion of the continents, c. sediments in the ocean, d. dating the atmosphere, e. dating the magnetic field

10. We must consider the possibility of processes happening at different rates. We can measure the rate that certain processes currently happen, but a massive flood or other event could have had a major impact in a very short amount of time.

11. The Bible says in Genesis 3 that the entire creation came under the curse of sin, including plants, animals, mankind, and the earth. As a result of the curse, everything is wearing down and deteriorating.

12. A majority of methods used to age-date the earth yield ages far less than the acclaimed billions of years.

13. Formed the cores of the continents; some erosion and deposition probably happened

14. The top of Mount Everest was once underwater and was later pushed up after the Flood waters receded.

15. The warm ocean waters rapidly evaporated and condensed over the colder continents, causing a buildup of ice and snow. See page 67 for more detailed information.

16. Review answer against student's previous essay on this subject.

## Unit Four Quiz, chapters 7–8

1. **Volcanism** — the eruption of molten rock (magma) onto the surface of the earth

2. **Escarpment** — a steep slope or long cliff that occurs from erosion or faulting and separates two relatively level areas of differing elevations

3. At the end of Noah's Flood it appears that a great volume of water was trapped, held in place by the Kaibab Upwarp. Ice Age rains filled the lake to overflowing, and as it burst through its mountain "dam," the huge volume of lake waters carved the canyon.

4. At Yellowstone Park, the soil and rock is thin, allowing very hot material to be near the surface. As rain and run-off water trickle down into the earth, they get heated, bubbling up in places as hot springs. In some places the underground water is trapped, and when heated to an excessive degree, it bursts out in a geyser.

5. The level of water in Lake Erie is somewhat higher than the elevation of nearby Lake Ontario. A river draining the waters of Lake Erie into Lake Ontario runs over the Niagara escarpment, resulting in a spectacular set of falls. Erosion takes place as the water roars over the falls, and the escarpment naturally recedes toward Lake Erie.

6. Both mountain chains are the result of layers of sediments deposited by Noah's Flood. The Appalachians buckled up in the early stages of the Flood and were subjected to massive erosion by the continuing Flood waters. The Rocky Mountains buckled up late in the Flood, extending up above the Flood waters as the waters drained off. Thus, the erosion to which they were subjected was much less intense.

7. Petrified wood can form, under laboratory conditions, in a very short period of time. The speed of petrifaction is related to the pressures that inject the hot silica-rich waters into the wood.

8. When water saturated with calcium carbonate enters an open space such as a cave, it cools off or evaporates, leaving the calcium carbonate behind. Stalactites (holding "tightly" to the ceiling) and stalagmites (which are usually larger and thus more "mighty" than stalactites) are formed as this calcium carbonate precipitates out of the water.

9. Coal can be formed in a laboratory by heating organic material away from oxygen but in the presence of volcanic clay. Under such conditions, coal can be formed in a matter of hours. One wonders if the abundant forest growing before the Flood would not have formed huge log mats floating on the Flood ocean. As these decayed and were buried by hot sediments in the presence of volcanic clay, they might have rapidly turned to coal.

10. Natural gas, mostly methane, is given off in the coalification process. The largest quantities of it, however, are found in deep rocks not associated with decomposition of organic material. Evidently, some natural gas is from both organic and inorganic sources.

11. Many theories have been promoted as to the specific origin of oil. The best seems to be that it is the remains of algae once floating in the ocean but buried in ocean sediments. Oil is not the remains of dinosaurs as has sometimes been claimed.

12. No — they are rare. Most fossils are of sea creatures, fish, and insects. Relatively few fossils are of land animals, specifically dinosaurs.

13. Answers will vary.

14. Answers will vary.

## *The Geology Book* ━● Test Answer Key

1. **Principle of uniformity** — the scientific thought that past processes are no different than processes today, meaning everything happens by gradual processes over very long periods of time

2. **Principle of catastrophe** — the scientific thought that sees evidence of rapid, highly energetic events over short periods of time, doing a lot of geologic work

3. **Erosion** — the process by which soil and rock are worn away

4. **Petrification** — the process by which trees, plants, and even animals are solidified by burial in hot, silica-rich water

5. **Turbidite** — an underwater, rapid deposition of mud that hardens into a layer of rock

6. **Gastrolith** — rounded stones used by plant-eating dinosaurs to aid in digestion and sometimes found with fossilized remains

7. **Fumaroles** — an opening in the earth's crust, usually associated with volcanic activity

8. **Metamorphism** — a process of heat and pressure that causes one rock to alter into another

9. **Magnetic field** — a field that exerts forces on objects made of magnetic materials; made up of many lines of force

10. **Sediment** — natural materials broken down by processes of erosion and weathering; can be transported or deposited by water or wind

11. **Second law of science** — also referred to as the second law of thermodynamics, which states that in every process or reaction in the universe, the components deteriorate

12. **Fountains of the deep** — a phrase mentioned in Genesis 7 as a reference to sources of water as part of the Great Flood of Noah

13. **Glacier** — a huge mass of ice that moves slowly over land

14. **Volcanism** — the eruption of molten rock (magma) onto the surface of the earth

15. Uniformity (the present is the key to the past) and catastrophe (highly energetic events operated over short periods of time and did much geologic work rapidly)

16. Crust, mantle, outer core, inner core

17. Creation, the Fall, Flood, Ice Age

18. The organism must be buried rapidly, protected from scavengers and from decomposition by bacteria and chemicals.

19. Whether it is soft or brittle, how deep it is buried

20. A majority of methods used to age-date the earth yield ages far less than the acclaimed billions of years.

21. Formed the cores of the continents; some erosion and deposition probably happened

22. The top of Mount Everest was once underwater and was later pushed up after the Flood waters receded.

23. The warm ocean waters rapidly evaporated and condensed over the colder continents, causing a buildup of ice and snow. See page 67 for more detailed information.

24. God sent the Flood as a judgment on the wickedness of mankind.

25. At the end of Noah's Flood it appears that a great volume of water was trapped, held in place by the Kaibab Upwarp. Ice Age rains filled the lake to overflowing, and as it burst through its mountain "dam," the huge volume of lake waters carved the canyon.

26. The level of water in Lake Erie is somewhat higher than the elevation of nearby Lake Ontario. A river draining the waters of Lake Erie into Lake Ontario runs over the Niagara escarpment, resulting in a spectacular set of falls. Erosion takes place as the water roars over the falls, and the escarpment naturally recedes toward Lake Erie.

27. Both mountain chains are the result of layers of sediments deposited by Noah's Flood. The Appalachians buckled up in the early stages of the Flood and were subjected to massive erosion by the continuing Flood waters. The Rocky Mountains buckled up late in the Flood, extending up above the Flood waters as the waters drained off. Thus, the erosion to which they were subjected was much less intense.

28. Petrified wood can form, under laboratory conditions, in a very short period of time. The speed of petrifaction is related to the pressures that inject the hot silica-rich waters into the wood.

29. a. Granite: Igneous, b. Marble: Metamorphic, c. Shale: Classic Sedimentary, d. Limestone: Organic Chemical Sedentary, e. Coal: Organic Chemical Sedentary, f. Rhyolite: Igneous, g. Slate: Metamorphic

30. First, layers of sediment are deposited. The weight of the water and the sediments on top begin to compact the sediments underneath. Next, warm water circulates throughout the sediments and dissolves certain minerals. The dissolved minerals surround the individual grains of sediment. Finally, when the water cools off and stops moving, the dissolved minerals act as a "glue" that cements the grains of sediment together to form sedimentary rock.

31. Heat and pressure recrystallize the minerals in rock into new mineral combinations. Some believe it happened over long periods of time; others believe it happened over short periods of time.

32. When a plant is living, it takes the isotope carbon 14 into its leaves, stems, and seeds. After the plant dies, the carbon 14 decays into nitrogen 14. Scientists can measure the amounts of both carbon 14 and carbon 12. Since they know the time it takes the isotope to decay, they can calculate when the plant died.

**Unit One Quiz, intro–chapter 1**

1. **karst** — the term used by scientists to describe a landscape of caverns, sinking streams, sinkholes, and a vast array of small-scale features all generated by the solution of the bedrock, formed predominantly by limestones

2. **Acheulean industry** — from the town of Saint-Acheul, whose most characteristic tool was the stone hand axe

3. **bas-reliefs** — artwork usually made of soft, pliable clay attached to walls or even to large blocks

4. **Kyr** — abbreviation for thousand years

5. **Myr** — abbreviation for million years

6. **Neanderthals** — believed by some to be early humans that were short, stocky, and stooped, with sloping foreheads, heavy eyebrows, jutting faces, and bent knees

7. **speleothems** — mineral deposits that form inside caves; especially stalagmites and stalactites

8. **karst aquifers** — the assembly of ground water accumulated inside a karstic rock, enough to supply wells and springs

9. Any two: cave bears, cave lions, cave hyenas

10. Paintings, engravings, and bas-reliefs

11. Their immediate need to find shelter from the rapidly cooling climate

12. It was deep inside the caves that some found shelter, mystical ritual hunting grounds, and a burial place for their dead.

13. 25 percent

14. Over 50 percent

15. Tower of Babel

16. The once-global knowledge and craftsmanship was split between many groups that could no longer truly communicate. Very quickly, various groups found themselves with the monopoly over one or several crafts/technologies, while other crafts were more or less lost for them. They were soon isolated from the other groups and many lost much of their knowledge of God.

17. It is first mentioned in Genesis 19:30 concerning Lot and his daughters.

18. China

19. (c) art associated with burial rituals

20. No, they were descended from the family of Noah.

21. The largest number of cave paintings are located in places of resonance (locations where if certain musical notes or sounds are emitted, they will bounce back, amplified, from the walls). It seems probable that chanting, dancing, and other types of ritual musical activities were associated with cave paintings.

**Unit Two Quiz, chapters 2–3**

1. **troglobites** — creatures that live only in caves (from Greek for "cave dwellers")

2. **troglophiles** — creatures that spend some part of their life in caves (from Greek for "who like caves")

3. **trogloxenes** — creatures that got into a cave by accident and try to leave (from Greek for "foreign to caves")

4. **endogenetic** — internal processes that can create caves

5. **exogenetic** — external process that can create caves

6. **resurgences** — the place where a sinking stream re-emerges to the surface

7. Two years; they were hiding from the Nazis.

8. Any four: Egypt, Phoenicia, Assyro-Babylonia, Greece, Rome, and Maya

9. Rock gypsum, rock salt, chalk

10. The cave olm

11. (b) troglophiles

12. A spectacular cave environment where several new species of creatures were found

13. Soluble rocks on which most landforms are formed by solution (karren, sinkholes, blind valleys, swallets, uvalas, poljes, etc.)

14. 12 percent

15. Named by Austrian geographers and geologists in the 19th century while studying limestone terrains; they Germanized the Slovenian name "Kras" used by the locals. Probably comes from an old pre-Indo-European root "kara" meaning "desert of stone."

16. Rhythmic springs flow intermittently, due to the very special shape and spatial distribution of the caves and conduits involved, as well as the constancy of water supply to the caves.

17. Usually around 90 percent

18.

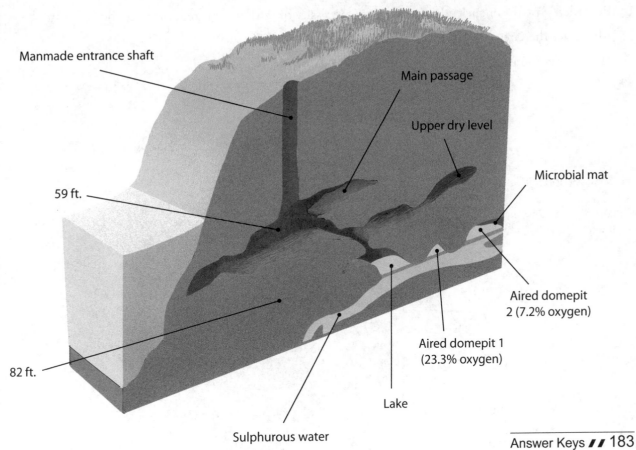

Manmade entrance shaft

Main passage

Upper dry level

Microbial mat

59 ft.

Aired domepit 2 (7.2% oxygen)

Aired domepit 1 (23.3% oxygen)

82 ft.

Lake

Sulphurous water

## Unit Three Quiz, chapters 4–5

1. **detrital formations** — sediments brought into the caves by streams and residual material left by the limestone

2. **dripping speleothems** — stalactites, stalagmites, and columns that are growing

3. **desiccation cracks** — cracks occurring because of shrinking of sediment as it dries

4. **master joints** — a tectonic discontinuity (fault line) that a given cave passage follows

5. Inflow, outflow, and through caves

6. Those that grow from stalactites (helictites) and those that grow from stalagmites (heligmites)

7. Any three: gypsum flowers, balloons, crusts, chandeliers, angel hair (mirabilite), moonmilk

8. Any four: good footwear (hiking boots or rubber boots), good rope, flashlights, hard hat, gloves, pack, wool socks, and knee pads

9. Ascending and descending by rope, crawling through muck and water, moving through large and small chambers, following master joints and scallops.

10. a. The Stermers (a Jewish family from Ukraine) b. Popowa Yama Cave

11. a. Humidity increases the absorption of light;

    b. unpleasant colors can result when using artificial light films;

    c. much time is required to take simple shots with artificial lights.

12. A special type of speleothem made up of two semicircular plates growing parallel to each other. They can grow more than three feet in diameter.

13. When cave pool water is saturated and very calm, thin flakes of calcite start growing, floating on the water, and are called cave rafts.

14. Recorded growths of speleothems refute the necessity of tens or hundreds of thousands of years.

15. Alexander the Great in 325 B.C.

16. Jaques-Yves Cousteau

17.

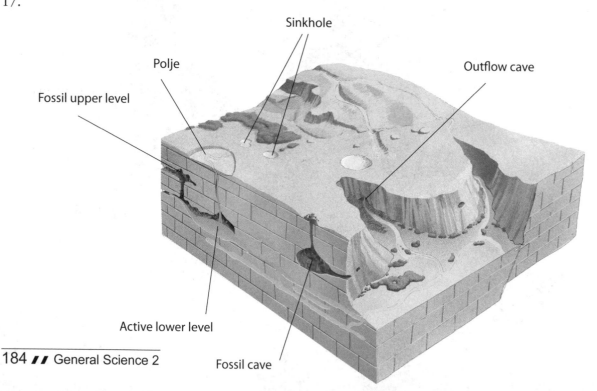

Sinkhole

Polje

Outflow cave

Fossil upper level

Active lower level

Fossil cave

**Unit Four Quiz, chapter 6**

1. **cenote** — steep-walled natural well reaching the water table and continuing below it

2. **concavities** — corresponding niches to convexities

3. **convexities** — a vertical succession of ledges

4. **diagenesis** — a complex set of transformations through which sediments go, from compaction, through dewatering to cementation

5. **Xibalba** — the Mayan "Underworld"

6. The precipitation water and the fluvial water

7. Hydrograph; chemograph

8. Subterranean geomorphology, geology, geochemistry, hydrology, hydrogeology

9. The karst denudation rate

10. As a transport and storage device

11. About 500 years

12. Water has a way of finding its way through karst aquifers, draining away from the artificial water reservoir.

13. What happens on the surface has a significant effect on what happens under the ground because infiltration from surface water can be extremely fast.

14. calcite

15. Answers may vary but should include the thought that it doesn't take millions of years for calcite to cover an item, only the right conditions.

16. Stage 1:     a: insoluble rocks

                      b: soluble rocks

                      c: hyperactive hydrothermal solutions generated during the Flood

                      d: large karst cavities excavated immediately after diagenesis

    Stage 2:     e: global tectonic uplift

                      f: global sheet flow

                      g: massive rain

                      h: new, detrital sediments

    Stage 3:     i: karstic sediments

1.  **bas-reliefs** — artwork usually made of soft, pliable clay attached to walls or even to large blocks
2.  **Kyr** — abbreviation for thousand years
3.  **Myr** — abbreviation for million years
4.  **troglobites** — creatures that live only in caves (from Greek for "cave dwellers")
5.  **troglophiles** — creatures that spend some part of their life in caves (from Greek for "who like caves")
6.  **trogloxenes** — creatures that got into a cave by accident and try to leave (from Greek for "foreign to caves")
7.  **detrital formations** — sediments brought into the caves by streams and residual material left by the limestone
8.  **dripping speleothems** — stalactites, stalagmites, and columns that are growing
9.  **desiccation cracks** — cracks occurring because of shrinking of sediment as it dries
10. **concavities** — corresponding niches to convexities
11. **convexities** — a vertical succession of ledges
12. **diagenesis** — a complex set of transformations through which sediments go, from compaction, through dewatering to cementation
13. Paintings, engravings, and bas-reliefs
14. rock gypsum, rock salt, chalk
15. Any four: good footwear (hiking boots or rubber boots), good rope, flashlights, hard hat, gloves, pack, wool socks, and knee pads
16. Hydrograph; chemograph
17. Tower of Babel
18. No, they were descended from the family of Noah.
19. (b) troglophiles
20. A spectacular cave environment where several new species of creatures were found
21. Recorded growths of speleothems refute the necessity of tens or hundreds of thousands of years.
22. Alexander the Great in 325 B.C.
23. About 500 years
24. Water has a way of finding its way through karst aquifers, draining away from the artificial water reservoir.

25.

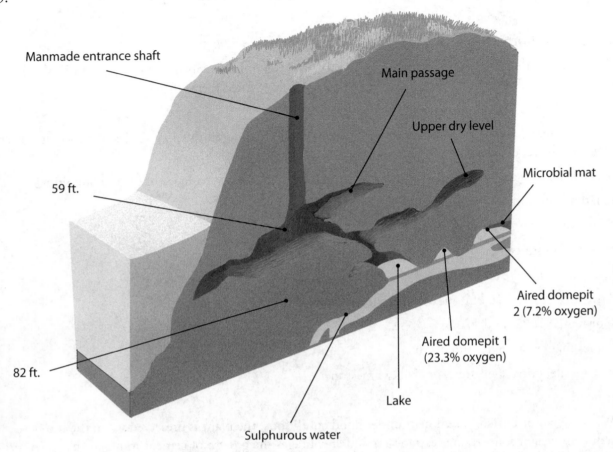

Manmade entrance shaft

Main passage

Upper dry level

Microbial mat

59 ft.

Aired domepit 2 (7.2% oxygen)

Aired domepit 1 (23.3% oxygen)

82 ft.

Lake

Sulphurous water

**Unit One Quiz, intro–chapter 1**

1. **Evolution** — the belief that life started by chance, and millions of years of struggle and death slowly changed a few simple living things into many complex and varied forms through stages

2. **Paleontologist** — a person who studies fossils

3. **Permineralized fossils** — fossils preserved by minerals hardening in the pore spaces of a specimen such as a shell, bone, or wood

4. **Trace fossils** — are not remains of plant or animal parts, but show evidence of once-living things

5. Charles Lyell and Charles Darwin

6. Wind and water; Water

7. Water and rock cement

8. Calcium carbonate and silica

9. Any three: limestone, bottom of tea kettle, in Tums and Rolaids, chalk

10. During the 1600s and 1700s

11. Time, chance, struggle, and death

12. Flaky shale, gritty sandstone, or chalky limestone

13. A flood

14. When a plant is buried in sediment under flood conditions, the plant is preserved when the heavy sediment weight squeezes out extra water and encourages the growth of cement minerals that turn the plant into a fossil.

15. The plant or animal needs to be preserved quickly before it begins to decompose.

16. A permineralized fossil

17. Permineralized wood has minerals in its pore spaces but still has wood fibers, while minerals have completely replaced the wood but preserved the pattern in petrified wood.

18. Coal is the charred remains and carbon atoms of once-living plants, making it a fossil. Coal burns, making it a fuel.

19. Huge mats of vegetation were ripped up in violent storms, torn apart by the waves and currents, and deposited in layers. Sediment on top of these layers then squeezed out water and raised the temperature of the buried plants. The plants would then begin to char, turning into coal.

20. The eruption of Mount St. Helens

21. If the layers surrounding the polystrate item had built up slowly over millions of years, the tops of the polystrate item would rot away, even if the bottoms were fossilized.

22. a. Creation b. God's perfect creation

23. a. Corruption b. Ruined by man's sin

24. a. Catastrophe b. Destroyed by Noah's Flood

25. a. Christ b. Restored to life in Christ

# Unit Two Quiz, chapters 2–3

1. **Index fossils** — fossils used to identify a geologic system because they lived either (a) at a certain time or (b) in a certain place in the pre-Flood world

2. **Geologic column** — a columnar diagram identifying rocky layers (strata) that form a sequence from bottom to top to indicate their relation to the twelve geologic systems

3. **Living fossils** — creatures found alive today that evolutionists thought became extinct millions of years ago

4. **Trilobite** — a crab-like creature that was the first fossil found buried in abundance around the world

5. **Cambrian explosion** — the sudden appearance of a wide variety of complex life forms in the lowest rock layer with abundant fossils (Cambrian); considered a challenge to evolution, these may be the first organisms in a corrupted creation to be buried in Noah's Flood

6. **Cavitations** — bubbles formed by surging waters

7. **Paraconformities** — a gap without erosion in the geologic column diagram; breaks the time sequence assumed by evolution, and may suggest fossils from different environments were rapidly buried by a lot of water, not a lot of time

8. **Stromatolites** — banded rock deposits formed by blue-greens growing in mossy mats on rocks in the tidal zone along the shore; the mats trap and then cement sand grains to form a mineral layer, continually building new layers on top of earlier ones

9. a. Sedimentary rocks (limestone, shale, sandstone); b. cliffs, cuts, creeks, and quarries

10. 12; 3

11. Charles Darwin; Charles Darwin realized that evolution needed viable evidence of transitions from one animal into another; without them, evolution could not be validated.

12. A series of burials

13. Since they were buried later in Noah's Flood, paleosystems with land plants and animals occur higher in the geologic column diagram than those with only sea creatures, but fossils of sea life occur in all geologic systems or eco-sedimentary zones since the Flood waters eventually covered all the land.

14. The waters that burst out of the deep during Noah's Flood

15. If the millions-of-years scenario were true and erosion occurred gradually, the softer rock would be gone and the hard rock would stick up into the sediment above. However, the tilted layers have been sheared off in a nearly straight line.

16. The bottom layers were most likely formed in the years before the Flood and sheared off during the beginning stages of the Flood. The upper layers were set into place during the Flood. Tectonic activity pushed these layers up as the water receded from the earth's surface during the later stages of the Flood. Water became trapped by earthen dams, which finally broke years after the Flood and released water to tear away the earth's surface. These cascades of water followed the easiest path downhill, which is now where we see the Colorado River through the gorge of Grand Canyon.

17. Mount St. Helens

18.

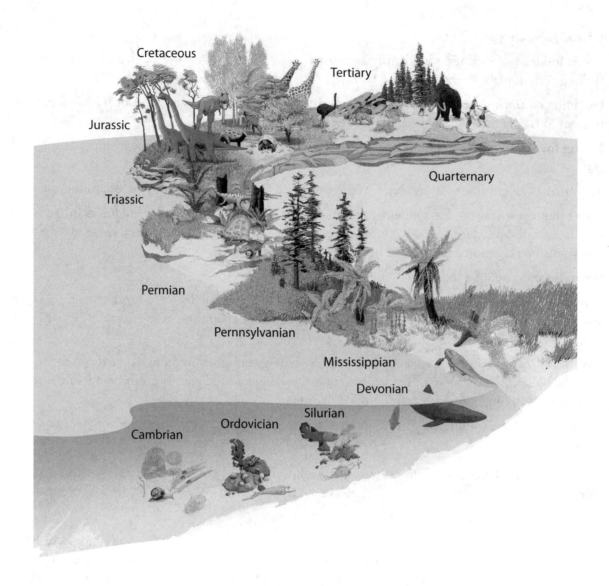

**Unit Three Quiz, chapter 4**

1. **Arthropod** — all creatures with jointed legs and a tough outside skeleton (exoskeleton) made of chitin: insects, crabs and shrimp, spiders, centipedes, and millipedes

2. **Cephalopods** — means "head-footed," since their tentacles come out of their heads; the most complex of all the invertebrates are the squid and octopus in the mollusk class

3. **Diatoms** — microscopic, one-celled plants whose walls are decorated with glass ($SiO2$) in exquisite patterns; mined and sold as diatomaceous earth, which is used in filtering and abrasion

4. **Echinoderms** — meaning "spiny-skinned," members of the starfish/sea star group usually have bony plates and spines embedded

5. **Gastropods** — means "stomach-footed," since they walk on their stomachs; mollusk class to which snails belong

6. **Malacology** — a branch of science devoted to the study of mollusk shells

7. **Nautiloids** — fossils with tapered, chambered shells; some are coiled like the modern nautilus, others are curved like bananas, and still others are straight, like ice cream cones

8. **Palynology** — the branch of paleontology that studies microscopic spores and pollen of plants

9. **Protozoan** — one-celled shelled animals

10. **Spicules** — sponges that have hard skeletal structures of crystal-like spines

11. a.  Trilobites

    b.  Evolution assumes that the earliest fossils, which are found in the lowest layers, would be the most primitive and least complex since they hadn't yet evolved into more complex beings. However, since these fossils reveal complex creatures of design, they disprove the idea that non-complex beings changed into complex beings.

12. Seashells

13. Many fossil clams are found with both sides of the shell still together. That means the clam must have been buried so deeply and so fast that it couldn't even open its shell to burrow out.

14. Shelled squids

15. Great Barrier Reef in Australia

16. Mississippian layer

17. They show that those areas were at one time completely covered by water.

18. It is possible that the volcanic activity that accompanied the Flood released toxins into the water that prevented the decomposition of the insects. Silt and clay buried the insects and settled quickly in briefly quiet water, solidifying fast enough to prevent later currents from tearing apart the fragile specimens. Insects can also be preserved in detail in hardened pine sap (amber).

19. The turbulent Flood waters covered the entire earth, including the high hills (Genesis 7:19). Then the mountains rose, and the valleys sank down (Psalm 104:8). At the end of the Flood, God raised up the layers that were below the sea, lifting sea-creature fossils even to the tops of earth's highest mountains.

20. Head

21. Eyes

22. Body (thorax)

23. Tail

24. Tail

25. Head with cheeks

26. Thorax

**Unit Four Quiz, chapter 5–conclusion**

1. **Evolutionary series** — a sequence of fossils that suggests how one kind of creature might have changed into another

2. **Metamorphosis** — the process of transformation from an immature form to an adult form in two or more distinct stages

3. **Splint bones** — modern one-toed horses actually keep parts of the two flanking toes as important leg support structures (not useless evolutionary leftovers)

4. **Vertebrates** — animals with backbones

5. Fish, amphibians, reptiles, birds, and mammals

6. The first explanation is because climate and soil conditions changed, dinosaurs had a difficult time surviving in that "new" world.

The second explanation is that they were over-hunted by people after the Flood. Both are certainly possibilities, but we are not absolutely certain why dinosaurs went extinct.

7.  a. amphibians

8.  Live coelacanths were found in the Indian Ocean and near Indonesia. There were no elbow or wrist joints as evolutionists once claimed; the stiff fin was used for steering and swimming, not walking. Their organs worked more like those in a shark, not those in a frog. The fish did not live in ponds; it lived in the deep ocean.

9.  God created all animals and people in the beginning to eat only plants. It was only after man's sin corrupted God's perfect creation that some animals began to eat other animals, and it was only after the Flood that God gave mankind permission to eat meat.

10. Dinosaurs were created on Day 6 along with the other land animals and people. Two representatives of the various dinosaur kinds were on the Ark. Since the average height of dinosaurs was about the size of a small pony, and since younger dinosaurs were smaller than older ones, they would have fit on Noah's Ark during the global Flood. Those that weren't on the Ark perished in the Flood. Many were buried in the muddy sediments. Those that survived the Flood on the Ark repopulated the earth after disembarking, although most eventually died from various causes.

11. 450 feet long; 75 feet wide; 45 feet high (150 x 25 x 15 meters)

12. They were likely young adults since God desired them to replenish the earth after the Flood.

13. Features in *Archaeopteryx* that evolutionists claimed to be transitional were and are found in other birds. The conclusion was made that *Archaeopteryx* was just a strong flying bird. Additionally, regular birds are found lower in the geologic column than *Archaeopteryx*.

14. First, the claimed earliest horse fossil is actually still seen today as a hyrax, or coney, not a primitive horse. Second, the change in horse size is not support for evolution since extreme sizes can be achieved simply by breeding. Third, the difference in hoof number can be explained by variation in horse kinds that were better able to live in different environments. Finally, the three-hoofed and one-hoofed horse kinds lived at the same time. These animals are not evidence for evolutionary transition.

15. The first buried fossils of each group are complete and complex, with all the features that separate its kind from all the others.

16. Since Adam sinned, the earth and all that was in it was cursed. Fossils themselves are dead things. Death was not part of God's original creation; it came as a result of sin. The fossil record also testifies to animals eating other animals; this did not occur before sin since there was no death before sin and since God originally created man and animals to be vegetarian.

17. See the seven explanations on pages 69–70.

    1.  Dead things are broken down so fast that most fossils must have formed rapidly or they wouldn't have formed at all.

    2.  Most fossils are found in sedimentary rocks that form in the way concrete cures, so the right conditions form rock quickly and no amount of time can form rocks under the wrong conditions.

    3.  Some dinosaur bones and other fossils contain DNA, protein, or other chemicals that would break down completely in thousands of years, not millions.

    4.  Countless numbers of living things must have been buried at the same time and place to form oil deposits, and that must have happened no more than thousands of years ago, or the oil would have leaked to the surface.

    5.  Gaps in the GCD with insufficient evidence of erosion, such as the "150 million missing years" in the walls of Grand Canyon, suggest evolution's millions of years are a myth.

6.  Misplaced fossils, like fossils of woody plants in Cambrian rock and living fossils, show that fossils from various geologic systems lived at the same time in different places, not at different times in the same place. The systems in the geologic column seem to be primarily the buried remains of different life zones in the pre-Flood world.

7.  Scientists studying the 1980 and 1982 eruptions of Mount St. Helens saw powerful evidence that catastrophic processes can do in days what slow processes could never do, not even in millions of years.

18. God preserved His creation while enacting judgment on the world. Noah, his wife, his sons, and their wives, along with two of every kind of unclean land animal and seven of every kind of clean land animal, were preserved by God's grace on the Ark. He promised judgment and yet made a way to escape that judgment. After the Flood, God restored His creation to life again. Man reproduced after his kind, animals reproduced after their kind, and plants reproduced after their kind, just as God commanded and desired.

## *The Fossil Book* ⟜⦿ Test Answer Key

1.  **Evolution** — the belief that life started by chance, and millions of years of struggle and death slowly changed a few simple living things into many complex and varied forms through stages

2.  **Paleontologist** — a person who studies fossils

3.  **Permineralized fossils** — fossils preserved by minerals hardening in the pore spaces of a specimen such as a shell, bone, or wood

4.  **Living fossils** — creatures found alive today that evolutionists thought became extinct millions of years ago

5.  **Trilobite** — a crab-like creature that was the first fossil found buried in abundance around the world

6.  **Cambrian explosion** — the sudden appearance of a wide variety of complex life forms in the lowest rock layer with abundant fossils (Cambrian); considered a challenge to evolution, these may be the first organisms in a corrupted creation to be buried in Noah's Flood

7.  **Arthropod** — all creatures with jointed legs and a tough outside skeleton (exoskeleton) made of chitin: insects, crabs and shrimp, spiders, centipedes, and millipedes

8.  **Cephalopods** — means "head-footed," since their tentacles come out of their heads; the most complex of all the invertebrates are the squid and octopus in the mollusk class

9.  **Diatoms** — microscopic, one-celled plants whose walls are decorated with glass ($SiO2$) in exquisite patterns; mined and sold as diatomaceous earth, which is used in filtering and abrasion

10. **Evolutionary series** — a sequence of fossils that suggests how one kind of creature might have changed into another

11. **Metamorphosis** — the process of transformation from an immature form to an adult form in two or more distinct stages

12. **Splint bones** — modern one-toed horses actually keep parts of the two flanking toes as important leg support structures (not useless evolutionary leftovers)

13. Charles Lyell and Charles Darwin

14. Charles Darwin; Charles Darwin realized that evolution needed viable evidence of transitions from one animal into another; without them, evolution could not be validated.

15. a. Trilobites

b. Evolution assumes that the earliest fossils, which are found in the lowest layers, would be the most primitive and least complex since they hadn't yet evolved into more complex beings. However, since these fossils reveal complex creatures of design, they disprove the idea that non-complex beings changed into complex beings.

16. The first explanation is because climate and soil conditions changed, dinosaurs had a difficult time surviving in that "new" world.
The second explanation is that they were over-hunted by people after the Flood. Both are certainly possibilities, but we are not absolutely certain why dinosaurs went extinct.

17. A flood

18. The eruption of Mount St. Helens

19. A series of burials

20. Since they were buried later in Noah's Flood, paleosystems with land plants and animals occur higher in the geologic column diagram than those with only sea creatures, but fossils of sea life occur in all geologic systems or eco-sedimentary zones since the Flood waters eventually covered all the land.

21. Many fossil clams are found with both sides of the shell still together. That means the clam must have been buried so deeply and so fast that it couldn't even open its shell to burrow out.

22. The turbulent Flood waters covered the entire earth, including the high hills (Genesis 7:19). Then the mountains rose, and the valleys sank down (Psalm 104:8). At the end of the Flood, God raised up the layers that were below the sea, lifting sea-creature fossils even to the tops of earth's highest mountains.

23. Features in *Archaeopteryx* that evolutionists claimed to be transitional were and are found in other birds. The conclusion was made that *Archaeopteryx* was just a strong flying bird. Additionally, regular birds are found lower in the geologic column than *Archaeopteryx*.

24. First, the claimed earliest horse fossil is actually still seen today as a hyrax, or coney, not a primitive horse. Second, the change in horse size is not support for evolution since extreme sizes can be achieved simply by breeding. Third, the difference in hoof number can be explained by variation in horse kinds that were better able to live in different environments. Finally, the three-hoofed and one-hoofed horse kinds lived at the same time. These animals are not evidence for evolutionary transition.

25. a. Creation b. God's perfect creation

26. a. Corruption b. Ruined by man's sin

27. a. Catastrophe b. Destroyed by Noah's Flood

28. a. Christ b. Restored to life in Christ

29.

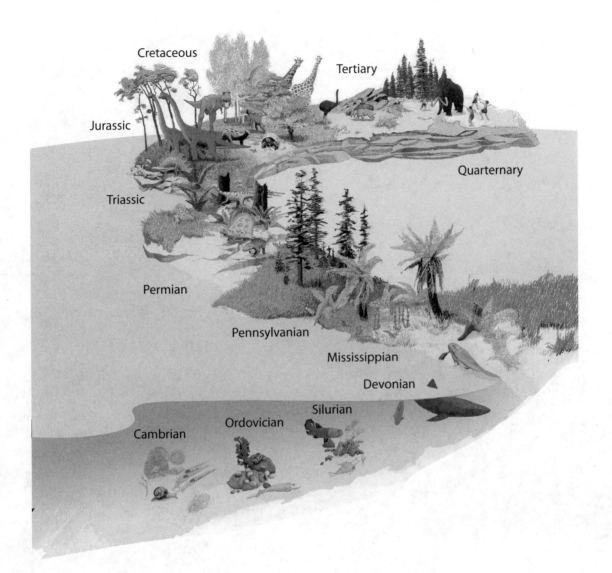

Cretaceous

Tertiary

Jurassic

Triassic

Quarternary

Permian

Pennsylvanian

Mississippian

Devonian

Silurian

Ordovician

Cambrian

**Accession year** — the year a king actually began his reign

**AD** — Anno Domini (the year of our Lord); the years after the Christian era began

**Amphitheater** — a circle of seats surrounding an area where gladiators fought each other or fought wild beasts

**Anatolia** — mountainous area in central Turkey

**Annunciation** — an announcement

**Archaeology** — the study of history — the history of the daily life of people and families and countries from hundreds and thousands of years ago

**Armenians** — people who live in eastern Turkey and northern Iraq

**Artifact** — an item from antiquity found in an excavation

**Asiatic** — in Egyptian terms, someone from Syria or Palestine

**Baal** — a word meaning "lord"; the name of a Phoenician god

**Bathhouse** — a club where citizens could bathe in cold, warm, or hot water

**BC** — Before Christ; the years before the Christian era began

**Baulk** — the vertical ridge left between two excavated squares in the ground

**Bedouin** — Arabs living in tents with no fixed address

**Bulla** — an impression made on clay with a seal (plural: bullae)

**Calvary** — Latin word meaning "skull"

**Carbon dating** — calculating the amount of carbon left in organic material that has died

**Causeway** — a built-up road

**Centurion** — a military officer in charge of a hundred men

**Ceramic** — something made of pottery

**Chaldees** — people who used to live in southern Iraq

**Chronology** — time periods, dates in which events happened

**Cistern** — a hole dug in rock to store rainwater

**Cuneiform** — a form of writing using a wedge-shaped stylus to make an impression on a clay tablet

**Debris** — discarded rubbish

**Dowry** — gift given to a prospective bride at the time of her marriage

**Drachma** — a Greek coin worth about a day's wages

**Dynasty** — a succession of kings descended from one another

**EB** — the Early Bronze Period

**Edom** — country in southern Jordan

**Edomites** — people descended from Edom, also known as Esau, Jacob's brother

**Exile** — a people sent out of their home country to another country

**Exodus** — going out; applied to the Israelites leaving Egypt

**Golgotha** — Hebrew word meaning "skull"

**Grotto** — cave

**Hieroglyphs** — Egyptian picture writing

**Inscription** — writing made on clay, stone, papyrus, or animal skins

**LB** — the Late Bronze period

**Mastabas** — mud-brick structures beneath which were tomb chambers

**MB** — the Middle Bronze Period

**Medes** — people who used to live in northern Iran

**Messiah** — meaning "Anointed One" and applied to an expected Jewish leader

**Millennium** — one thousand years

**Nabataeans** — people descended from Nabaioth who occupied Petra

**Nomad** — a person who lived in a tent that could be moved from place to place

**Non-accession year** — the first complete year of a king's reign

**Nubia** — a country south of Egypt now called Sudan

**Ossuary** — a box in which human bones were preserved

**Papyrus** — sheets of writing material made from the Egyptian papyrus plant

**Passover** — Jewish ceremony celebrating the Exodus from Egypt

**Persia** — a country in central Iran

**Pharaoh** — title applied to many Egyptian kings

**Pottery** — a vessel made of clay fired in a kiln

**Rhyton** — a drinking vessel shaped like a human or animal

**Sarcophagus** — a stone coffin

**Scarab** — model of a dung beetle with an inscription engraved on it for sealing documents

**Seal** — an object made of stone, metal, or clay with a name engraved on it used to impress in soft clay

**Scroll** — papyrus or animal skin document rolled up into a cylinder

**Siq** — narrow valley between two high rock formations

**Stratum** — a layer of occupation exposed by excavations (plural: strata)

**Synchronism** — something happening at the same time

**Syncline** — a boat-shaped geological formation

**Tell** — a Hebrew word meaning "ruins"; applied to hills on which people once lived

**Theater** — a stage for actors in front of which was a semi-circle of seats

**Vellum** — animal skin treated to be used as writing material

**Wadi** — a dry river bed, carrying water only when it rained

**Yehovah** — a Hebrew name for God, usually spelled Jehovah, but there is no "J" in the Hebrew alphabet

# The Geology Book - Glossary

**Alluvial sediment** — material carried by fast-moving rivers and streams that are deposited at points where the water moves slower

**Asthenosphere** — a suspected area in the uppermost portion of the earth's mantle where material is hot and deforms easily

**Atom** — the basic component of chemical elements

**Basalt** — a type of igneous rock that makes up most of the oceanic crust; on land it forms when extruded by volcanoes or through fissures

**Canyon** — large areas with steep walls that have been carved out of layers of sedimentary rock

**Carbon dating** — a process that uses the decay of carbon 14 to estimate the age of things that were once living

**Catastrophism** — the philosophy about the past, which allows for totally different processes and/or different process rates, scales, and intensities than those operating today; includes the idea that processes such as creation and dynamic global flooding have shaped the entire planet

**Cavitation** — bubbles within fast moving water explode inwardly and pound against a rock, eventually reducing to powder

**Cementation** — a process in the formation of sedimentary rock when minerals are dissolved which then help to solidify the rock

**Chemical rock** — a type of sedimentary rock built up as chemicals in water, usually seawater, precipitate and consolidate

**Clastic rock** — a type of sedimentary rock consisting of fragments of a previously existing rock (e.g., sandstone consists of consolidated sand grains)

**Concretions** — concreted masses of sedimentary rock that has been eroded out of a softer area of rock

**Continental separation** — the concept that the continental plates have moved apart (or collided), concluding, for example, that Africa and South America were once connected

**Continental shield** — the primarily granite core of a continent that has been exposed to the surface and then bulges up because there is no weight upon it

**Compaction** — a process in the formation of sedimentary rock when the materials are pushed together tightly, leaving little to no open spaces

**Coprolite** — fossilized animal or dinosaur dung; can be used to determine a creature's diet

**Core** — the center of the earth is thought to be a sphere of iron and nickel, divided into two zones; the outer core is in molten or liquid form, while the inner core is solid

**Crossbed** — areas of extremely large ripple marks

**Crust** — the thin covering of planet Earth, which includes the continents and ocean basins; nowhere is it more than 60 miles (100 km) thick

**Decomposition** — the process by which things are broken down into smaller or more basic substances or elements

**Deposition** — the process by which sediments are deposited onto a landform

**Earthquake** — a sudden release of energy below the earth's crust, which causes the earth's crust to move or shake

**Erosion** — the process by which soil and rock are worn away

**Escarpment** — a steep slope or long cliff that occurs from erosion or faulting and separates two relatively level areas of differing elevations

**Fault** — a fracture in rock along which separation or movement has taken place

**Fold** — a bend or flexure in a layer of rock

**Fossil** — the direct or indirect remains of an animal or plant

**Fountains of the deep** — a phrase mentioned in Genesis 7 as a reference to sources of water as part of the Great Flood of Noah; some biblical scholars feel it could refer to oceanic or subterranean sources of water

**Fumaroles** — an opening in the earth's crust, usually associated with volcanic activity

**Gastrolith** — rounded stones used by plant-eating dinosaurs to aid in digestion and sometimes found with fossilized remains

**Geyser** — underground water that has been heated to an excessive degree and because of pressure, bursts out of the ground temporarily

**Glacier** — a large natural formation of ice where the accumulation of ice and snow exceeds the amount it melts or turns from a solid to a gas

**Granite** — a widespread igneous rock, which contains abundant quartz and feldspar, and makes up a significant portion of the continental crust

**Igneous rock** — rock formed when hot, molten magma cools and solidifies

**Isotope** — variations of an element's atoms, usually in the different number of neutrons

**Kolk** — an underwater tornado that lifts or removes underlying materials

**Law of disintegration** — states if processes continue as they do today, everything will eventually be eroded or worn away

**Magnetic field** — a field that exerts forces on objects made of magnetic materials; made up of many lines of force

**Mantle** — beneath the thin crust and above the core of the earth; it is about 1,864 miles (3,000 km) thick

**Metamorphism** — a process of heat and pressure that causes one rock to alter into another

**Metamorphic rock** — rocks formed when heat, pressure, and/or chemical action alters previously existing rock

**Mountain** — a large landform rising abruptly from the surrounding area

**Obsidian** — a common type of rhyolitic volcanic rock, which almost looks like a chunk of black glass

**Petrification** — the process by which trees, plants, and even animals are solidified by burial in hot, silica-rich water

**Plain** — a broad area of relatively flat land

**Plate** — the earth's crust, both continental and oceanic, is divided into plates, with boundaries identified by zones of earthquake activity; the idea of plate tectonics holds that these plates move relative to one another, sometimes separating or colliding, and sometimes moving past each other

**Plateau** — flat lying sediment layers similar to plains but at higher elevations

**Polar ice cap** — a high latitude region of a planet that is covered in ice

**Principle of uniformity** — the scientific thought that past proccesses are no different than processes today, meaning everything happens by gradual processes over very long periods of time

**Principle of catastrophe** — the scientific thought that sees evidence of rapid, highly energetic events over short periods of time, doing a lot of geologic work

**Radioisotope dating** — the attempt to determine a rock's age by measuring the ratio of radioactive isotopes and the rate at which they decay

**Rhyolite** — molten rock that forms granite that has erupted on land and solidified

**Ripple marks** — marks that indicate moving water flowed over a rock layer when the sediments were still muddy and yet to harden

**Second law of science** — also referred to as the second law of thermodynamics, which states that the entropy (disorder) of the universe increases over time

**Sediment** — natural materials broken down by processes of erosion and weathering; can be transported or deposited by water or wind

**Sedimentary rock** — rock formed by the deposition and consolidation of loose particles of sediment, and those formed by precipitation from water

**Sedimentation** — the act of depositing sediments

**Tsunami** — often called a tidal wave; a seismic sea wave produced by an underwater disturbance such as an earthquake, volcano, or landslide; can be extremely destructive

**Turbidite** — an underwater, rapid deposition of mud that hardens into a layer of rock

**Uniformitarianism** — the philosophy about the past that assumes no past events of a different nature than those possible today, and/or operating at rates, scales and intensities far greater than those operating today; the slogan "the present is the key to the past" characterizes this idea

**Uplift** — in geology, a tectonic uplift is a geological process most often caused by plate tectonics, which causes an increase in elevation

**Volcanism** — this is the process by which molten rock or lava erupts through the surface of a planet

**Acoustics** — points of resonance (locations where if certain musical notes are emitted, they will bounce back, amplified, from the walls)

**Acheulean industry** — from the town of Saint-Acheul, whose most characteristic tool was the stone hand axe

**Active caves** — live caves that have a flowing stream in them

**Arthropods** — invertebrate animals having an exoskeleton, segmented body, and jointed appendages

**Bas-reliefs** — artwork usually made of soft, pliable clay attached to walls or even to large blocks

**Bidirectional air circulation** — air flowing two ways

**Cave** — considered a natural opening in rocks, accessible to humans, which is longer than it is deep and is at least 33 feet in length

**Cave paintings** — either simple outlines of charcoal or mineral pigment, or true paintings with outlines, shading, and vivid pigment fills

**Cenote** — steep-walled natural well reaching the water table and continuing below it

**Compoundrelict caves** — fossil caves above the water table

**Concavities** — corresponding niches to convexities

**Convexities** — a vertical succession of ledges

**Cul-de-sac** — cave with only one entrance

**Denudation rate** — the pace at which a given surface of bare rock is eroded, usually measured in millimeters per millennium (thousand years)

**Desiccation cracks** — cracks occurring because of shrinking of sediment as it dries

**Detrital formations** — sediments brought into the caves by streams and residual material left by the limestone

**Diagenesis** — a complex set of transformations through which sediments go, from compaction, through dewatering to cementation

**Dripping speleothems** — stalactites, stalagmites, and columns that are growing

**Echolocation** — bats send out sound waves that hit an object and an echo comes back, helping them identify the object

**Emergences** — the place where subterranean waters emerge to the surface

**Endogenetic** — internal processes that can create caves

**Engravings** — usually made on soft limestone surfaces

**Exogenetic** — external process that can create caves

**Karst** — the term used by scientists to describe a landscape of caverns, sinking streams, sinkholes, and a vast array of small-scale features all generated by the solution of the bedrock, formed predominantly by limestones

**Karst aquifer** — the assembly of ground water accumulated inside a karstic rock, enough to supply wells and springs

**Karsted** — rich in karst features, especially caves

**Kyr** — abbreviation for thousand years

**Master joint** — a tectonic discontinuity (fault line) that a given cave passage follows

**Myr** — abbreviation for million years

**Neanderthals** — believed by some to be early humans that were short, stocky, and stooped, with sloping foreheads, heavy eyebrows, jutting faces, and bent knees

**Orthokarst** — karst formed on carbonate rocks mainly by solution

**Parakarst** — karst-like features formed on non-carbonate rocks, mainly by solution

**Phreatic caves** — flooded (water saturated) caves formed and/or located below the water table

**Pseudokarst** — karst-like features formed on any kind of rock by other ways than solution

**Relict caves** — caves without a flowing stream, which may have ponds or dripping water

**Resurgences** — the place where a sinking stream re-emerges to the surface

**Scallops** — spoon-shaped hollows in a cave wall, floor, or ceiling dissolved by eddies in flowing water

**Sinkholes** — funnel-shaped hollows, from a few feet to hundreds of feet in diameter

**Speleothems** — mineral deposits that form inside caves; especially stalagmites and stalactites

**Troglobites** — creatures that live only in caves (from Greek for "cave dwellers")

**Troglophiles** — creatures that spend some part of their life in caves (from Greek for "who like caves")

**Trogloxenes** — creatures that got into a cave by accident and try to leave (from Greek for "foreign to caves")

**Unidirectional air circulation** — air flowing one way

**Vadose caves** — caves that formed and continue to exist mostly above the water table; the majority of their passages have air above the water

**Xibalba** — the Mayan "Underworld"

**4 C's** — an aid for remembering four major events in biblical history important to understanding fossils: God's perfect Creation, ruined by man's sin (Corruption), destroyed by Noah's Flood (Catastrophe), restored to new life in Christ

**Archaeology** — the science that deals with human artifacts, and with things deliberately buried by humans

**Arthropod** — all creatures with jointed legs and a tough outside skeleton (exoskeleton) made of chitin: insects, crabs and shrimp, spiders, centipedes, and millipedes

**Artifacts** — products crafted by humans

**Cambrian explosion** — the sudden appearance of a wide variety of complex life forms in the lowest rock layer with abundant fossils (Cambrian); considered a challenge to evolution, these may be the first organisms in a corrupted creation to be buried in Noah's Flood

**Cavitations** — bubbles formed by surging waters

**Cephalopod** — means "head-footed," since their tentacles come out of their heads, the most complex of all the invertebrates are the squid and octopus in the mollusk class

**Creationist** — one who thinks that (1) fossils show complex and separate beginnings because each kind was created well designed to multiply after kind, but that (2) fossils also show death, disease, and decline in variety and size because struggle and death followed man's sin (until Christ returns) and brought on Noah's Flood

**Diatoms** — microscopic, one-celled plants whose walls are decorated with glass ($SiO_2$) in exquisite patterns; mined and sold as diatomaceous earth, which is used in filtering and abrasion

**Echinoderms** — means "spiny-skinned," members of the starfish/sea star group usually have bony plates and spines embedded

**Evolution** — the belief that life started by chance, and millions of years of struggle and death slowly changed a few simple living things into many varied and complex forms through stages

**Evolutionary series** — a sequence of fossils that suggests how one kind of creature might have changed into others

**Evolutionist** — one who believes fossils will show that (1) millions of years of time, chance, struggle, and death changed a few simple life forms into all the complex and varied forms we have today, and that (2) new structures gradually developing from low to high in the geologic column will be seen when "missing links" are eventually found

**Fossil** — remains or trace of a once-living thing preserved by natural processes, most often by rapid, deep burial in waterlaid sediments

**Gastropods** — means "stomach-footed," since they walk on their stomachs; mollusk class to which snails belong

**Geologic column** — a columnar diagram identifying rocky layers (strata) that form a sequence from bottom to top to indicate their relation to the twelve geologic systems

**Geologic system** — a major rock layer whose fossils are used to name it for one of the twelve groups in the geologic column diagram

**Geology** — the scientific study of the earth, including the materials that it is made of, the physical and chemical processes that occur on its surface and in its interior, and the history of the planet and its life forms

**Index fossil** — fossils used to identify a geologic system because they lived either (a) at a certain time or (b) in a certain place in the pre-Flood world

**Invertebrate** — animals without backbones

**Living fossils** — creatures found alive today that evolutionists thought became extinct millions of years ago

**Malacology** — a branch of science devoted to the study of mollusk shells

**Metamorphosis** — the process of transformation from an immature form to an adult form in two or more distinct stages

**Mollusks** — a large phylum of animals with thick, muscular bodies and a complex system of organs

**Nautiloids** — fossils with tapered, chambered shells; some are coiled like the modern nautilus, others are curved like bananas, and still others are straight, like ice cream cones

**Paleontologist** — a person who studies fossils

**Paleontology** — the study of fossils

**Palynology** — the branch of paleontology that studies microscopic spores and pollen of plants

**Paraconformities** — a gap without erosion in the geologic column diagram; breaks the time sequence assumed by evolution, and may suggest fossils from different environments were rapidly buried by a lot of water, not a lot of time

**Permineralized fossils** — fossils preserved by minerals hardening in the pore spaces of a specimen such as a shell, bone, or wood

**Petrified** — fossils preserved by minerals completely replacing but preserving the pattern in the original wood, bone, etc.

**Polystrates** — fossils that cut through many layers, suggesting the sequence was laid down very rapidly

**Protozoan** — one-celled animal

**Pseudofossils** — false fossils; things that look like fossils but really aren't

**Sediments** — particles of sand, silt, clay, ash, etc., eroded and deposited by wind and water currents

**Spicules** — sponges that have hard skeletal structures of crystal-like spines

**Splint bones** — modern one-toed horses actually keep parts of the two flanking toes as important leg support structures (not useless evolutionary leftovers)

**Stratigraphic series** — sequence of fossils from lower to higher in the geologic column diagram (see above); thought to represent either (a) stages in evolution, or (b) stages in burial during Noah's Flood

**Stromatolites** — banded rock deposits formed by blue-greens growing in mossy mats on rocks in the tidal zone along the shore; the mats trap and then cement sand grains to form a mineral layer, continually building new layers on top of earlier ones

**Trace fossils** — are not remains of plant or animal parts, but show evidence of once-living things

**Trilobite** — a crab-like creature that was the first fossil found buried in abundance around the world

**Vertebrates** — animals with backbones

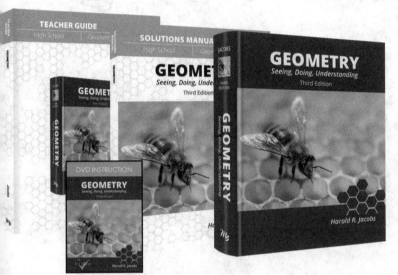

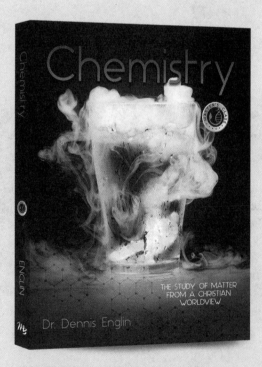